Additional copies of "Irrevy" are avail[able from the Committee] for Nuclear Responsibility, Main P.O.[...] CA 94101.

"*Irrevy*"

An IRREVERENT, ILLUSTRATED VIEW of NUCLEAR POWER

John W. Gofman

A collection of talks.

From Blunderland to Seabrook IV

Published by the
Committee for Nuclear Responsibility
Main P.O.B. 11207
San Francisco, CA 94101.

CNR is a non-profit educational organization.

This collection of talks is published as a public service. Neither the author nor the artists are receiving any royalties. The author receives no remuneration of any sort from the Committee. The artists asked for only nominal fees or none at all; their generosity in no way means that they necessarily agree with all the ideas in these talks. Indeed, it would be surprising if the Committee's own board members were to agree with every idea herein, for they are each well known as *independent* thinkers too.

Distributed by
THE CROSSING PRESS
Trumansburg, New York 14886

 First edition.

Library of Congress Card Catalog #78-73212

International Standard Book Number 0-932682-00-6

This book is dedicated
to the cartoonists,
whose insights contribute so enormously,
so delightfully,
to public education and human decency.

■ Contents ■

■ ■ ■ ■ ■

Index by Issues

These talks deal with a great many issues. The chance that there exists a single reader who will agree with *every* point in the book, is (to borrow a classic line from the nuclear sales-force) about as low as the chance that you or I will be struck by a meteor. But these talks are *meant* to provoke discussion and independent thinking. Feedback from distribution of the individual talks indicates that they *do* provoke a lot of thought—which is why it seemed like a useful project to put them together as a book.

It may surprise you to find an index in the *front* of this book, but I think you may find it useful here because it is by *issues*. There is a strictly alphabetical index, mostly for names, in the back.

JWG.

■ NUCLEAR POWER ■

Nuclear Power, continued

Nuclear Power, continued

■ HEALTH EFFECTS OF EXPOSURE TO RADIATION ■

■ ENERGY-EFFICIENCY ■

■ SOLAR ENERGY ■

■ PUBLIC HEALTH CONCEPTS ■

■ TECHNOLOGY ■

■ SCIENCE AND MEDICINE ■

■ INALIENABLE, NATURAL, HUMAN RIGHTS ■

■ LIBERTY AND SLAVERY ■

JUSTICE, MORALITY, AND THE LAW

NUCLEAR WAR AND NUCLEAR WEAPONS

■ MANKIND'S WORST PROBLEMS: ■ Confronting the Source

■ SUGGESTIONS FOR POSSIBLY USEFUL ACTIONS ■

□ □ □ □ □

It is a privilege to acknowledge the seminal ideas of Lewis Mumford concerning many of the issues discussed in this book. *JWG.*

Special thanks to Egan O'Connor, whose ideas, values, reasoning, and writing contributed enormously to the content of these talks. In a very real sense, the authorship of this book is as much hers as mine. *JWG.*

We are particularly grateful to Maisie McAdoo, who did almost all the typesetting and who provided first-rate guidance on additional matters. She is not responsible, however, for our somewhat unorthodox use of commas to reflect speaking rhythm. We thank What's Your Line Graphics (a San Francisco typesetting co-op) for special helpfulness too.

It gives us great pleasure also to thank Ramesh Patel, the Xerox wizard of United Copy Service (San Francisco) for his most generous and talented pre-printing help.

Good advice and goodwill were extended to us in the pre-printing stage also by Carl Gosline of Postal Instant Press, Bob Barton of Michael's Artist Supplies, and by Jim Folk and John Keeling of Blue Print Service Company (all of San Francisco).

The printing was done by Consolidated Printers (Berkeley), where it was a pleasure to work with "can-do" people like Richard T. Pisani, Rudy Torres, and Frank Consul.

Last but not least, we thank Ena Macrae of the *San Francisco Chronicle* for her help in locating so many of the cartoonists. The saying comes to mind: You can easily judge the character of a person by how he treats those who can do nothing for him.

□ □ □ □ □

ACCIDENT

As "Irrevy" goes to press, the nuclear power accident at Three Mile Island (PA) is still going on. Almost everything we would say about the accident is already in "Irrevy" except for this:

According to the Nuclear Regulatory Commission, the dose to the public so far is 1,800 "person-rems"...a figure which we regard with extreme skepticism because it turns out there existed almost no monitoring system, and because that estimated dose excludes a significant additional internal dose from the radioactive Carbon-14 released.

A person-rem is the dose (in rems) per person, times the number of people who receive it (or, rems times people). A rem is a unit of energy from radiation; there are, of course, 1000 millirems in a rem, so one millirem can be written 0.001 rem. It is the energy of ionizing radiation which breaks important molecules in our cells. Most injured cells will not go on to become cancers, but some will.

For every 300 person-rems of whole-body irradiation delivered to a population (adults and children mixed), one person has been condemned to die later from cancer or leukemia. (At least one death per 300 person-rems).

Suppose an exposure-rate is "only" 2 millirems (0.002 rems) per hour per person, but exposure continues for 3 days (72 hours) and it reaches 1,000,000 people. You multiply the people by the rems to get person-rems, and you divide by 300 to estimate the number who will die later from the irradiation (the rems, hours, and persons "cancel out"):

$$\left(\frac{0.002 \text{ rems}}{\text{hour}}\right) \times \left(72 \text{ hours}\right) \times \left(1{,}000{,}000 \text{ persons}\right) \times \left(\frac{1 \text{ death}}{300 \text{ person-rems}}\right)$$

$$= 480 \text{ deaths.}$$

It is a fraud to imply that a dose "equivalent to just a few chest X-rays" to a million people would be safe. If you give low-dose radiation to a lot of people (whether from nukes, X-rays, or Mother Nature), you will kill just as many people as if you gave high-dose radiation to just a few people, providing the number of person-rems is the same. It's not hard to understand, if you remember that 6 x 2 = 12, but 2 x 6 also = 12.

JWG *JWG.* April 4, 1979

1.

ALICE IN BLUNDERLAND

Phil Frank. College Media. Berkeley CA.

The issue of acceptability of nuclear power as a significant energy source for the USA divides itself logically into considerations of the near- and mid-terms (up to 1985-1990), and separately for the long term (beyond 1990).

For the near- and mid-term, the question is whether the light water reactor (on which the U.S. program is almost wholly based) is acceptable or helpful for our energy problem. For the period beyond 1990, the light water reactor would appear to have no future in any event, so the question then becomes the acceptability of "breeding" either with the plutonium-based liquid-metal fast breeder reactor or one of the combined uranium-thorium cycles, with uranium-233 as the key fissionable nuclide. In my opinion the reasons for rejection of both nuclear programs are overwhelming, which I hope to demonstrate in the ensuing remarks.

Energy: Buy It and Throw It Away?

There is simply no doubt that a supply of energy is needed, and a *reliable* supply, if a modern "developed" society is to continue to function. It is *not* my purpose to examine here whether our quality of life is, or is not, ideal. It would be rather arrogant for anyone to decide what represents a good quality of life. I shall, for present purposes, make the assumption that the American standard of living, the life-style of most people with their amenities of life, are to remain at current levels.

It is virtually self-evident that if energy supply goes down, or if its dollar price goes up, then the standard of living will necessarily fall *provided* all else is equal. However, in a most important way, all else need not be equal. In our energy usage, there is a fraction which is sheer waste. The *wasted* energy contributes absolutely nothing to anyone's standard of living. It therefore follows, to the extent that unnecessary waste can be eliminated, that our standard of living can even go *up* in the face of a decline in energy sources or rise in prices. In every important sector of the U.S. economy where energy is used—home, industry, agriculture, commerce, and transportation—the energy *waste* varies from high to profligate, all without one iota of contribution to the amenities of life which people have chosen to enjoy.

It is unfortunate that two *very* separate concepts have been commonly lumped together, with enormous obfuscation of the energy problem (generally this is intentional and self-serving). The first concept is life-style change, including less driving of autos, less illumination, a lower setting of the home thermostat plus wearing of sweaters, etc. The second, and truly important, concept is that of ENERGY EFFICIENCY, which means doing the same task without *throwing away* our energy supply!

For example, a gas tank in our car with a hole in it, losing gasoline, hardly contributes to our standard of living. It would be energy-efficient to plug the hole. Most of our energy inefficiencies are more subtle than this, but they are no less wasteful and hence destructive to our standard of living.

Curiously, the promoters of nuclear power almost invariably mention only the first concept. For a long time it has been said, "An appeal to patriotism is the last refuge of the scoundrel." In our modern society, the last refuge of the promoter is to threaten people with loss of jobs, loss of livelihood, lack of food, and a return to life in a cave. I am very disappointed that 32 well known scientists recently resorted to this tactic in their brazen promotion of nuclear power for the USA.

In a country which, through inefficiency, is wasting 45% of the energy it "consumes", it is simply dishonest for anyone to claim that banning additional nuclear power plants would interfere with economic growth and employment. It should be self-evident that any country which is using its energy with only 55% efficiency, could almost *double* its economy without needing more power plants of *any* sort.

The USA uses energy with only 55% the efficiency with which Sweden and West Germany use it. So I propose a simple policy: The Monkey-See, Monkey-Do Plan. Let's hold our annual energy consumption steady at its present level, and increase our *efficiency* by 3% per year.

Energy efficiency, as a concept in conservation, is likely to become a great stimulus to employment, as well as *the* truly significant way to approach the energy supply problem in the near- and mid-term. The question that needs

to be asked is, "What is best for our economy? An energy-efficiency strategy that will provide jobs and save enormous quantities of energy? Or a light water reactor program (providing many fewer jobs per capital dollar invested) so that we may continue to throw our energy away?"

There is much said and written these days about a "capital crunch," a shortage of available capital to do all the things which society would like to have done. If, indeed, there is a shortage of capital, then it is especially important to use the capital in those endeavors which are most effective for the economy. Investment in energy efficiency is far more attractive than investment in *wasting energy*, which is what choosing the light water reactor program represents!

Capital, *per se*, could hardly care less *what* it is invested in, since the only criteria are safety of principal coupled with the highest rate-of-return per capital dollar invested. Hula hoops, nuclear reactors, or energy-efficient buildings are equally acceptable to capital which is seeking return. Therefore, it is up to society to insist upon where the capital should go. For those who fear that capital will be withheld unless it can be invested in *anti*-societal endeavors, there is no need for fear. Capital will not sit idle, since the whole game is to use capital for its appreciation.

Benefits from Fixing Our Buildings

The American Institute of Architects has recently issued two important, practical, and carefully researched reports (1) which describe a program for energy-efficient buildings which, if instituted now and built up over the next 15 years, would result in a savings, from this source alone, of some 12 million barrels of oil per day in 1990 and thereafter. Moreover, the program would have an *early* yield of energy saved—some 750,000 barrels of oil per day after the first year.

Our total energy consumption in the USA is currently in the neighborhood of 70 quads per year (one "quad" represents one quadrillion BTUs, or British Thermal Units), and is predicted to rise to something on the order of 100 quads by 1990. The saving of 12,000,000 barrels of oil per day means making available 26.3 quads of thermal energy

+ + + + + + + + + + + + + +

(1) *Energy and the Built Environment* (1974) and *A Nation of Energy Efficient Buildings by 1990*, from the American Institute of Architects, 1735 New York Ave., N.W., Washington, D.C. 20006.

per year. THIS IS SOME 37% OF CURRENT TOTAL ENERGY USE, AND WOULD BE 26% OF THE ABOVE-PROJECTED 1990 USE OF ENERGY! So, energy-efficient buildings could provide a very large contribution to our energy requirement, and do so while providing a large boost to the economy and to employment. The obvious advantages of early implementation of the energy-efficient-buildings strategy are:

(1.) The technology is known; no breakthroughs are required.

(2.) There would be a greater energy-yield per capital dollar invested, compared to investing the same capital to build nuclear plants for the purpose of wasting energy.

(3.) The energy-yield per dollar spent by the consumer would be greater than for nuclear power.

(4.) High energy-return per capital dollar and per consumer dollar supports a healthy economy, a good living standard, and jobs.

(5.) The number of jobs provided (in the retrofitting of existing buildings with energy-efficient systems, and installing energy efficiency in new buildings) should be much greater than an equivalent capital investment in nuclear plants.

(6.) The AIA estimates of 30% energy savings on retrofitted buildings, and 60% energy savings on new buildings, are regarded as *conservative* estimates.

(7.) The Energy Research and Development Administration (ERDA) agrees that the energy-efficiency strategy must be regarded as a prime factor in national planning for solution of our energy problem.

(8.) This establishment of a whole, new, burgeoning industry to stop wasting energy can begin to pay off in barrels of oil saved NOW rather than at some indefinite future time.

(9.) If the concept of energy efficiency is put into vigorous operation in the building sector, successes will stimulate related efforts in transportation and industry, both areas where it is widely admitted that enormous waste occurs.

It seems difficult to believe it could make sense to invest capital and job effort in nuclear power FOR THE PURPOSE OF WASTING ENERGY YEAR AFTER YEAR, when the capital and job effort can be spent in energy-efficiency with effects lasting into the long-range future.

20th Century N-Power = Trivial Power

NET energy can be won from the current light water reactors *if* there exists a guaranteed supply of rich uranium ores worth working. Yet the current situation is that light water reactors are being currently built and planned, for which the uranium supply is exceedingly uncertain. The supply of ores of a practical grade is poorly known. We may quote ERDA itself on this: "Estimates of recoverable uranium resources currently contain a high degree of uncertainty." ERDA adds that with a vigorous search program, the resources might become better known by 1980.

Does it make sense to try to solve our midterm problem by dependence upon an *insecure* source of fuel? The uranium supply for the current light water reactor technology is estimated to be 500,000 tons of uranium oxide in high grade ore reasonably assured domestically and, by great optimism, ERDA estimates that number *might* go as high as 3,600,000 tons. Under ideal conditions, with vastly improved performance over the current dismal performance of the light water reactors, these estimated reserves could be worth some 190 to 1,380 quads (thermal) to our energy economy. Since these reserves would be used over a 30 year period, the potential gain to our energy budget is between 6.3 and 46

Courtesy of William H. Gruen, *Critical Mass Journal*.

quads per year, the latter figure representing supreme optimism. Probably a value between 10 and 15 quads per year is tempered with some realism, and this is only some 10% to 15% of our estimated 1990 energy consumption.

It appears that, even with optimism, the light water reactor program cannot assure over half the energy contribution projected for the energy-efficiency strategy in buildings. The uncertainty in usable uranium ores is such that the nuclear promoters may be cornered in the quandary that reactors are being planned and *sold* WHICH MAY NOT BE REFUELABLE. The effect of this fiasco would be:

(1.) To hit the electric rate-payer with an enormous additional cost;

(2.) To create real havoc for the country's economy, through industry having counted upon power from this source; and

(3.) To cause a loss of jobs, under such circumstances.

What do we hear from the nuclear promoters to respond to these issues? One answer that defies all rationality is that we can purchase uranium ores from foreign sources! These promoters seem to have forgotten the whole basis for energy concern, namely, our experience with the OPEC producers. They are suggesting we get held up by *two* cartels instead of one. A second answer, which has been seriously proposed, is to use Tennessee Shale of which we have mountain loads.

The BIG difficulty with this is that such shale is 60 parts per million uranium oxide, versus 2,000 ppm for good grade ores. The energy required to win the uranium from low grade ores will go up as the ore grade goes down. Chapman (2) has estimated this energy requirement will be inversely proportional to the ore grade in ppm of uranium oxide. This seems very reasonable to me, and based upon such an estimate, I would predict that Tennessee Shale will use more energy in being won than it can provide in current light water reactors. Morgan Huntington (3) has repeatedly made this point before. Certainly no credible authority has demonstrated evidence upon which the conclusion can be supported that such shale can be used to support the viability of the current light water reactor program. Even a confirmed nuclear super-optimist, Dr. Ken Davis, a Vice-President of Bechtel Engineering Co., has recently written the following for the Atomic Industrial Forum (1974):

> "Assuming 85% recovery of uranium from Chattanooga Shale, then the rock as mined has the same fuel-value as the best grades of coal, pound for pound, if the recovered uranium is used to fuel the types of reactors being built and operated today."

One strains hard to believe that this is actually the

\+ + + + + + + + + + + + + +

(2) "The Ins and Outs of Nuclear Power", by Dr. Peter Chapman, in *The New Scientist*, December 19, 1974.

(3) Huntington is an engineer; he testified about Tennessee Shale and other matters as an expert witness before the Nuclear Regulatory Commission, May 20, 1975.

statement of a nuclear power expert from one of the largest design engineering firms in the nuclear power business. The vaunted industry which shows you a *handful* of uranium pellets on the TV screen, and equates it with a *carload* of coal, is now down to admitting we may be using ore that requires a carload of shale for a carload of coal. Quite a comedown! To go on from Dr. Davis' absurd recommendation that we dig up a good part of Tennessee for this trivial yield per pound, we must point out the ridiculous position it represents, once the additional energy to isolate the uranium oxide is considered. If someone starts with rock no more energy-rich than coal, as mined, he might even be in energy *debt* by the time the uranium is used in today's light water reactors.

Unless a really reliable source of high-grade uranium ore is assured to fuel each reactor, it may well be appropriate to use Huntington's description of the promotion and sale of such reactors as the "biggest swindle of modern times." Of course the electric rate-payer, and the people who lose their jobs, will be the victims of this swindle, since utilities are guaranteed their fair return on investments.

Summary: Near- and Mid-Term N-Power

Thus, if we should decide to invest hard-to-get capital in the light water reactor program, WE MAY REAP:

(1.) The burden of a major white elephant on our hands in the power-delivery business: unfuelable nuclear power stations.

(2.) A situation in which industry really does not have the power it was expecting to have; the economic dislocation would make the current recession and joblessness look very mild.

(3.) A very poor energy-yield for the economy per dollar of capital invested, compared with energy-efficiency(4); and the comparison will be even more dismal if the present poor performance of nuclear plants continues with respect even to *energy-output*.

(4.) A huge economic dislocation, if a single major accident occurs at one of the plants, and all the other plants have to be shut down *suddenly*, without time to provide for substitute-energy.

All this, for WHAT? All for a *possible*, small yield of energy — 10% to 15% of our annual energy requirement! Current nuclear technology looks like an exceedingly poor investment of scarce capital dollars when contrasted with investment in energy-efficiency. This is true in reliability, in jobs created, in health of our economy. OUR TOP PRIORITY FOR INVESTMENT SHOULD BE IN ENERGY-EFFICIENCY, NOT IN NUCLEAR POWER.

Long-Term Options: Breeder vs. Solar

The proposed breeder reactors, of whatever type, could (in theory) truly yield large quantities of net energy, even from poor fuel-sources such as Tennessee Shale, simply because a very much larger *fraction* of the energy in the uranium would be utilized. At this moment, it is unknown whether or not breeder reactors can be made to *work* successfully. On other grounds, namely health and safety, there are very good reasons to reject the breeder reactors without committing further massive waste of public and private funds. But before considering the health-evidence itself, I must state the reasons why there is no need to go

\+ + + + + + + + + + + + + +

(4) In April 1977, Jimmy Carter's *National Energy Plan* stated that "Conservation is cheaper than the production of new energy supplies", and provided figures showing it to be 2-fold to 6-fold cheaper than oil. In the *Wall Street Journal* of February 9, 1978, "Firms Spend Millions to Cut Energy Use; Payoff Is Often Rapid", Ralph E. Winter reports: ". . . Energy conservation, say corporate money men who keep an eye on high fuel and electricity costs, is about the most profitable use for capital these days . . ."

forward with a concept so chancy and hazardous as the breeder:

(1.) The breeder reactor cannot possibly be of any help in contributing energy before the period 1990 or 2000. ERDA recently speaks of beginning contributions by the year 2000.

(2.) Much more attractive alternatives definitely exist.

Let me quote from a recent statement from ERDA(5): "Solar energy falling on about 3% of land, if utilized at about 10% efficiency, could meet the total projected U.S. energy requirements for the year 2000." Explicitly including solar energy "among the inexhaustible resources to be given high priority," ERDA stated, "The technologies for producing essentially inexhaustible supplies of electric power from solar energy will be given priority comparable to fusion and breeder reactors." [It is the author's opinion that the serious disadvantages of *fusion* power deserve wider recognition, too. *]

So . . . finally . . . after four years of denial by various authorities—particularly by the incredibly promotional Atomic Energy Commission—solar energy is now being accorded recognition as a reasonable option for solving our long-term energy needs. ERDA may inadvertently have given us the real reason why solar energy was made to *look* so enormously difficult or impossible, with the following statement: "Transitions to new systems must occur without major disruption of existing systems. Existing investments must be paid for, and represent an inertial force on the system."

How true! $100 billion dollars are on the line, and the nuclear industry plus its governmental handmaidens

\+ + + + + + + + + + + + + +

(5) *A National Plan for Energy RD&D,* Report 48, June 28, 1975, by ERDA, Washington, D.C. 20545.

* *Power from Fusion: The Meaning of Recent Breakthroughs,* a C.N.R. report, August 15, 1978.

By permission of Community Press Features, Boston 02116.

plan to shove nuclear power down America's throat, "to recover the investment," no matter what the cost in *health* to you or your children.

Health and Safety: A Credibility Issue

Now let us turn our attention to the health and safety aspects of nuclear power. Particularly let us examine what, if anything, can be believed out of the nuclear establishment.

(1.) The "Safe Dose" Fraud

Up to 1969 the AEC and the nuclear industry, with elaborate public relations campaigns, *promoted* the idea that below a certain level, radiation would do no harm to humans. There never was, and there is not now, *any* evidence for a "safe" amount of radiation. *Every* responsible organization studying radiation injury now holds that cancer, leukemia, and genetic damage must be considered to be essentially proportional to dose, down to the very lowest radiation doses. What can you trust from an industry and bureaucracy which actively promote such a vicious, harmful doctrine to the public?

(2.) How Many Extra Cancer Deaths?

In 1969 Dr. Tamplin and I said, in a scientific paper, that the existing so-called "safe" standards for public exposure would cause 32,000 extra cancer deaths per year, *if* the public were exposed to the legal limit. The genetic consequences, after several generations, could be between 100,000 and 1,000,000 extra deaths per year.

For the ensuing two years, the AEC and the nuclear industry attempted ridicule and denial. Indeed, AEC and industry scientists vied with each other for who could show we were wrong by the largest factor. So the numbers went from "wrong by 100 times," "wrong by 1,000 times," to even "wrong by 10,000 times." Sadly for their credibility, many of the enthusiastic apologists put their statements in print.

Finally, a committee of the National Academy of Sciences completed a two-year study of the question and reported, among other findings, that

(1.) We were correct that no evidence at all exists for a safe amount of radiation;

(2.) We might be 4 to 10 times too high in our cancer estimates. In the fine print of the report, the

Committee admitted that all the evidence was not in, and that it might have to raise its estimates towards ours. *

What are you to believe from an industry and bureaucracy which talk of 0 to 3 cancers per year, from a dose that the National Academy of Sciences admits will cost *many thousands* of extra cancer deaths per year?

(3.) Actions vs. Words

Faced with total discrediting by the National Academy of Sciences report, the AEC and the nuclear industry retreated to a new position. They shouted loudly, "We'll never give you the dose permitted by the regulations. In fact we'll never give you more than one two-hundredth of the dose permitted by the regulations. We won't give you more than one millirem per year from the entire nuclear power industry, even after it grows to the size expected in the year 2000."

How *very* attractive. When innumerable people suggested lowering the standards for allowable exposure, the industry fought this tooth and nail—and they still fight it today. The industry spokesmen say repeatedly, "Why lower the permissible dose, when we don't plan to give you that dose anyhow."

Why? Because the history of polluters is that they pollute as much, or even MORE, than is legal! When an industry fights *hard* to prevent reduction in a poison's legal limit, it is fair to infer it is because that industry plans to give us *at least* the presently permitted dose! Actions speak louder than words. Resistance to lowering the permissible dose speaks more loudly than cheap promises.

\+ + + + + + + + + + + + + +

* The scientific reports in which we presented our findings are referenced at the back of this book. The report of the National Academy of Sciences, by its Advisory Committee on the Biological Effects of Ionizing Radiations, is *The Effects on Populations of Exposure to Low Levels of Ionizing Radiation*, November 1972; it is known also as the B.E.I.R. Report.

It is true that the nuclear industry picked out one miniscule portion of the places and ways it can irradiate you—namely, from the power plant itself under so-called "*normal* operating conditions"—and the industry agreed to lower the allowable dose from that miniscule portion of the chain. Isn't that charitable of it?

(4.) Catch-22: No Dose-Limit!

Incredible as it may seem, some of the scientific supporters of the nuclear industry babble utter nonsense about radiation exposure. For example: "You can stand next to a nuclear power plant and get *less* radiation than you do from the radioactive potassium in your body." Some of them are even more cute, and they say, "Your wife's potassium-40 will give you more radiation in bed than you'd get standing next to a nuclear power plant."

Probably today every third grader understands that, if you put enough steel and concrete between yourself and a radiation source, you'll get very little radiation. Certainly when a nuclear reactor is operating perfectly, and there is enough steel and concrete, of course there is a very low dose at that point. But let us suppose the reactor is NOT operating perfectly. Will they still promise those same beautifully low doses? "Oh no," say the industry and the government bureaucracy, "under those circumstances, we are to be given a variance for unplanned events." So any doses delivered ABOVE those promised simply don't count at all, because they fall into the category of "unplanned" or "abnormal" circumstances!

What are you to think of an industry and bureaucracy which set up a perfect Catch-22, to be allowed to give you ANY dose they give you—depending upon how well or poorly their systems operate—by the simple expedient of calling all over-doses "unplanned"?

Editorial cartoon by Bill Sanders.
Reprinted courtesy of Field Newspaper Syndicate.

(5.) Natural and Medical Irradiation

Next, the promoters say, "Why worry about nuclear power? We live in a sea of radioactivity. Even Mom's apple pie is radioactive. The average exposure from natural radiation sources is 100 millirems per year, and man has come a long way in spite of *that* exposure throughout his history. Why worry about a small addition to that 100 millirems?"

Let's consider natural radiation for a moment. A dose of 100 millirems is that dose. So the anticipated costs in the USA of natural radiation are estimated to be approximately 19,000 extra cancer deaths per year, and between 58,000 and 580,000 genetic deaths per year. But it isn't easy to move from the Earth! So we must suffer these tragic consequences of natural radiation. It would represent sheer idiocy to ADD unnecessary deaths from cancer and genetic disease to these already high numbers. If anything, the effect of natural radiation is so bad that we should do everything possible to avoid making it worse.

Next, the advocates of nuclear power say, "If you're worried about all those people being killed by radiation, why do you tolerate unnecessary medical radiation, which amounts to half or two-thirds of natural radiation?" On this point I must say I find myself in total agreement with the nuclear industry. Medical radiation can and must be *reduced*! It is NOT self-evident that nuclear medicine is really saving more people than it is killing. Put another way, if you have enough medical exams, you'll get sick.

(6.) Averaging Some Stolen Years

The industry's next argument is that any deaths which will occur from nuclear power, will steal only a few minutes or hours from your life-span. Isn't it worth a few minutes or hours to enjoy abundant electric power?

How very reasonable. Let me explain how this estimate is arrived at. Suppose a person is victimized by nuclear power radiation, and dies at age 20 of a radiation-induced cancer. If everyone in the population makes a loan of some seconds, then the victim's loss of life-span is drastically reduced. The only difficulty is that this particular banking transaction or averaging can't be worked out, so the 20-year-old victim has still lost 45 years or so of life-span.

What can you believe from an industry which puts forth such misleading arguments?

(7.) Silence about Genetic Injury

Let us now examine some of those promised small doses, and look at what is really one of the most pernicious frauds on health yet perpetrated by the nuclear industry. No one disagrees that genetic injury from ionizing radiation is a most serious consequence; as a result, numerous geneticists have recommended that the nuclear power industry never be permitted to deliver more than *one* percent of the dose received from natural sources. The National Academy of Sciences Report advised limitation to a *few* percent of natural radiation, and the EPA (Environmental Protection Agency) has recently concurred in this advice. Why worry, says the nuclear industry, we never plan to deliver more than *one* percent of natural radiation.

But what the nuclear industry does not highlight, or even mention, is that ONE ROUTE of genetic exposure to the public will in all likelihood deliver some *eight times* this much radiation, even if everything *else* goes perfectly. From published AEC reports, the data are available for the radiation exposure received by occupational workers in the nuclear industy. As the nuclear plants have increased in size, this exposure to workers has increased, and there is a possibility it may increase even further as the plants age. Even if it does *not* increase per plant, and if we accept the projected number of plants for the year 2000, then the effect of such occupational exposure will be the *same* as giving the ENTIRE POPULATION approximately *eight* percent of the natural radiation dose—a dose eight times as high as *promised* by the nuclear industry.

From the viewpoint of genetic injury, the human species will suffer *just as much* if the genetic damage is produced in fertile occupational workers, as if the same amount of radiation dose were introduced into the population *as a whole*. But the public is unaware of this fact of life concerning genetic damage, and the nuclear industry isn't about to inform the public. Instead, the nuclear industry, and the government bureaucracies which support it, simply *ignore* the occupational exposure, when they calculate population exposure! They refuse to consider workers in the nuclear industry as part of the population—a simply marvelous ruse.

So clever are they with this device, that when they have a particularly "dirty" task to perform, and when the regular workers are already at the legal exposure limit, they bring in workers (as many as 2,000 for a single job) to work for a few minutes or hours, during which they may get three months of legal exposure. But this exposure doesn't count as public exposure because, by *their* definition, "Any person inside the boundary of the plant is no longer a member of the population." Incredible as it may seem, the EPA supports this "definition."

Certainly these facts must magnify your confidence in the nuclear industry and in the government bureaucracy which supports it.

(8.) No Test = No Harm?

Another, common, fraudulent argument of the nuclear industry relates to the occurrence of cancer in regions which vary from each other in natural radiation levels. There are places in the world (such as Denver) where the radiation exposure from natural sources is twice as high as the average. Some places are even higher. The nuclear industry is fond of saying, "The people residing in such areas don't suffer from the extra radiation that they receive." Such a statement is an absolute, unequivocal lie. What the nuclear bureaucracy *means* is that no valid scientific study has ever LOOKED for such injury! It does *not* mean the injury is not occurring. For the nuclear industry and its advocates, no *test* is the same as no *harm*. Credibility?

(9.) No Little Flags = No Victims?

Another, favorite, nuclear cliche is "The nuclear industry is so safe that *not a single* radiation casualty has yet occurred." Another scientifically fraudulent statement, on several separate grounds.

The nuclear industry is terribly prone to forget crucial parts of the entire nuclear power cycle. The cycle starts with uranium mining. One hundred uranium miners are already dead of radiation-induced lung cancer. The estimates are that another 1,000 may die of exposure they have *already* had, even if they stop uranium mining now. Among radiation workers at nuclear plants, the doses *already* received and documented are going to result in hundreds of cancers, even though the symptoms have not yet appeared.

These people have already had their radiation death-warrant sealed. How do the bureaucrats treat these deaths? Very simple! They just deny them. All medical scientists know that once a cancer occurs, it is impossible to say which of several possible cancer-provoking agents caused that particular case. The cancers don't sprout A LITTLE FLAG announcing which agent caused the cancer. So the nuclear industry just says, "Prove that radiation caused *this* case!" Thus, it perpetrates the outrageous not-a-single-death deception. But this is not a new technique. Even though medicine knows that cigarette smoking causes some 90% of lung cancers, the cigarette companies have not lost any lawsuits yet. They simply say, "Prove that *our* cigarettes caused this case."

Lastly, it is known and documented that radiation-releases are *planned* and occur *regularly* at operating nuclear facilities, and that population exposure *occurs* as a result. Since there is NO dose of radiation safe with respect to cancer-production, it follows that the number of cancer deaths caused by this so-called "safe" industry is directly proportional to the dose it has deliberately delivered.

This doesn't begin to include all the accidental spills and *un*measured releases of radioactivity. In this industry, the fox guards the chicken coop! The nuclear industry *itself* tells us how much radioactivity it has released. On every occasion where an independent measurement was made, the nuclear industry was caught short!

Is this an industry you can trust?

(10.) Miracles Promised

It is important to examine how well the nuclear industry must contain its monstrous radioactive garbage, to avoid giving the population quite *high* doses. The containment must be at least 99.9% perfect, under routine AND non-routine operating circumstances combined. I leave it to you to decide whether you believe it will accomplish this particular miracle, among the many miracles its advocates are promising.

Plutonium Toxicity and Containment

If the nuclear industry is to have any *viability* beyond the short term, one or another of the various "breeding" cycles must become operational to provide nuclear fuel. One such cycle is the LMFBR (liquid metal fast breeder reactor), which means society faces a plutonium future (or lack thereof). Another cycle is the thorium-uranium cycle, which means society must face plutonium *plus* uranium-232 and uranium-233. The choice between plutonium on the one hand, and the uranium nuclides on the other, is comparable to the choice between death from cyanide or death from hemlock. For plutonium, the deaths will be from lung cancer. For the thorium cycle, with its uranium-232 and uranium-233 products, the toxicity will be comparable, but the cancers will be in different organs (rather than primarily in the lung), and in addition, there will be more *genetic* injury.

A most lovely choice.

Just how toxic is plutonium? Dr. Dixy Lee Ray, ex-Chairperson of the now defunct Atomic Energy Commission, travels widely, insisting that botulism toxin is more poisonous than plutonium, weight for weight. She *may* be correct in that happy-go-lucky insistence. But what would you THINK of anyone planning to build our energy-future on something like botulism toxin, assuming energy *could* be derived from it?

I've recently published two papers on the production of lung-cancer by plutonium. For the plutonium produced by power reactors, I estimate about 21 billion lung-cancer doses per pound, if it is finely divided into small particles and if all the particles become trapped in human lungs. (This order of toxicity would be the case for plutonium's distribution into a population of half cigarette smokers, half non-smokers—the smokers being much more susceptible.) I think my estimate may be two times too high, but it is about equally likely that my estimate may be two times too low.

"Aha," says the nuclear industry, "that's *if* plutonium gets inhaled, but how will it get inhaled?"

"Aha," says the nuclear industry, "we've seen five tons of plutonium distributed all over the globe by weapons-testing fallout, and we don't know of a single death caused by plutonium inhalation!"

Is that really so? I am prepared to defend, before any scientific body, and under oath in full public view, my estimate that ONE MILLION people (perhaps only 500,000 or as many as two million) in the Northern Hemisphere have been irreversibly condemned to die of lung cancer from those five tons of plutonium. Indeed, were it not for the fact that by far *most* of the plutonium fell either upon the oceans or uninhabited land, the figure of one million would be enormously larger.

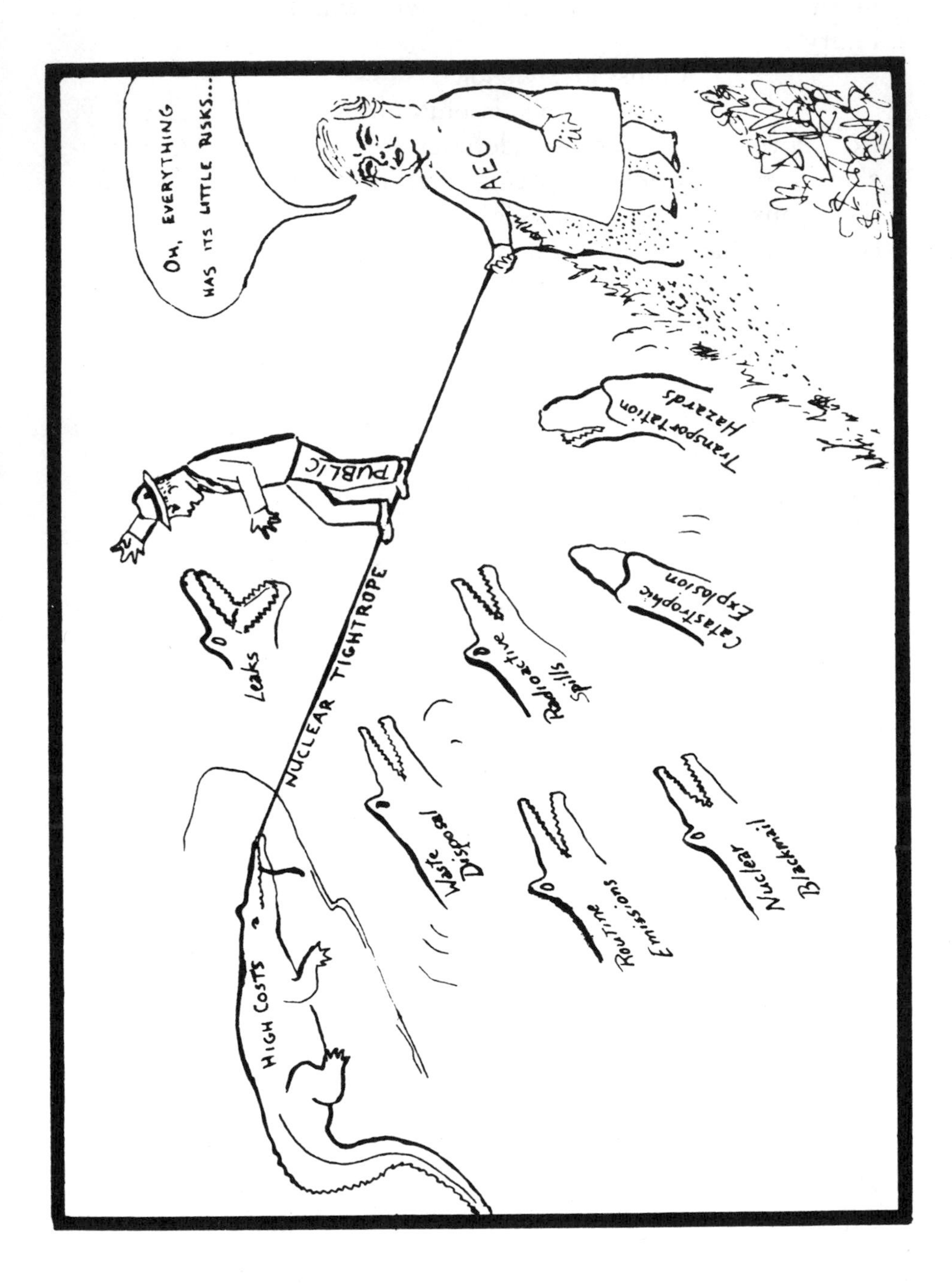

By Janet Kailin, Sequim WA 98382.

Yet the very same people who used to tell you that nuclear weapons-testing in the atmosphere was "safe", when it really meant a million deaths from lung cancer due to plutonium, are *now* telling you they will put 440 *million* pounds of plutonium through the nuclear cycle SAFELY in a full plutonium breeder economy! Let's now ask how well they must do, to make good on that promise.

If they *contain* the plutonium well enough to prevent 99.99% from getting airborne (quite a feat), it will *still mean* about 500,000 additional fatal lung-cancers per year in the United States alone, from the tiny fraction which *does* get airborne. A half-million extra cancers will represent some six times the current death-rate from lung-cancer from all causes combined.

I've debated this point with some nuclear experts, and they have one stock answer: "We'll prevent any more than one part in a billion from getting out." If that answer does not instantly strain your credulity, let me give you some facts.

Some *Real* Plutonium Experiences

At Rocky Flats (Colorado), plutonium is handled in quantities of hundreds of kilograms for the fabrication of nuclear weapons. One kilogram is the equivalent of 2.2 pounds. Let's look at the Rocky Flatters' record. They do many of the same operations which will be done in a plutonium *energy* economy, plus some additional work on plutonium metal required for weapons fabrication.

• In 1969 they had a fire which finally was estimated to have cost $45,000,000—the largest industrial fire in United States history. Why did they have this fire? Because the heat-sensors, for a glove-box containing plutonium metal, were placed in a location where they were *well insulated* from the heat of spontaneously burning plutonium! As a result, the fire was raging by the time the alarms sounded. (A small oversight, said management after the fire.)

• Colorado environmentalists asked whether any appreciable amount of plutonium had been released offsite by the fire.

• "No," said AEC and the Dow management, "our air samplers show that very little escaped beyond the boundary of the site."

• So Dr. Edward Martell of the Colorado Citizens' Environmental Committee made measurements, and found that about ONE HALF POUND of plutonium was on the ground east of the Rocky Flats plant. Remember, *if* all of it got trapped in human lungs, a half pound would cause billions of lung-cancers. It's deadly, and it's not all on the *ground* forever. The winds there will put some of it back in the air for people to breathe. Furthermore, a half-pound is what is on the ground nearby, and doesn't even count whatever was carried in the air to Denver and beyond.

• First, Dr. Martell's measurements were ridiculed. Then several independent sources, plus AEC's Health and Safety Laboratories, confirmed that Dr. Martell was certainly in the *right* ball-park about how much was on the ground outside the plant.

• Then came the real frosting on the cake. The AEC insisted that this very large amount of plutonium which had escaped, really didn't come from the fire. Instead, it came from the wind stirring up plutonium that had leaked out of some 5,500 storage-barrels which were sitting around *rusting* on the Plant site. Although they knew by 1962 that the barrels were leaking, they continued to put more plutonium-containing barrels out to rust for an additional six years. And to save money, they used *old* barrels. I think it was in 1972 that someone there suggested it might be wise to store wastes in *new* barrels.

• There have been innumerable additional spills at Rocky Flats, both inside and outside buildings. This experience is by no means unique to Rocky Flats. I would urge you to read Robert Gillette's classic articles in *Science* magazine about plutonium spills, about plutonium tracked out by workers, about plutonium *accidentally discovered* on the ground and elsewhere in numerous plants handling plutonium, and finally, about some plutonium which found its way to a restaurant in one instance, and to the local sheriff's office in another!

Future for the Digging Industry

To borrow a recent phrase from Hubert Humphrey, I would say that the credibility of the nuclear industry's ability to control plutonium to one part in a billion is about equal to that of a $3 Confederate Bill made in Germany. The Dow Company managers of Rocky Flats have been forever digging up parts of the Rocky Flats site, and shipping plutonium-contaminated soil to Idaho, to bury it there. They estimated in a recent report that, with their current ability to make shipments to Idaho, they could ship the contaminated soil out of Rocky Flats to Idaho within 349 years.

After we spilled plutonium in Spain, we spent over 50 million dollars digging up a good part of the countryside and bringing the contaminated soil back to South Carolina for burial. At the rate the nuclear industry messes up on a *small* scale, it will be digging up half the United States to bury on the *other* half, in a full-fledged plutonium economy.

Protection by "Health" Physicists

"But," says the nuclear industry, "we are very concerned about your health. We hire large numbers of health physicists to protect the public." Let me give you three illustrations of the marvelous protection which *that* provides.

SUPPORT NUCLEAR POWER! TODAY'S TECHNOLOGY MAKES IT SAFE
CHLORINE
AUTH

(1.) *Taking Up "Residence"*

At Rocky Flats recently, as a result of much adverse publicity about the sloppy handling of plutonium, a health physicist introduced a marvelous solution to the plutonium problem there. He suggested they *re-define* the word, "contamination". Contamination, he suggested, should only be used when the spill is above some arbitrarily defined level. All spills below that level are to be described either as "A plutonium infiltration has occurred", or as "Plutonium has taken up residence at Site XYZ". As a result of this miraculous solution of the containment problem, plutonium contamination has declined dramatically at Rocky Flats. What wondrous marvels the English language can accomplish.

(2.) ". . . *Where Your Money Is*"

Dr. Dade Moeller, in his inaugural presidential address to the society of health physicists, was describing the opportunities for health physicists in the burgeoning nuclear industry. In discussing the hazards of radiation with the health physicists, he exhorted his confreres to become active nuclear advocates with these exact words: "Put your mouth where your money is."

(3.) *Only One Chance in Thirty!*

Recently the State of New York discovered that air shipments of multi-kilogram quantities of plutonium-oxide were being made into Kennedy Airport. This particulate form of plutonium-oxide is the worst conceivable form for lung-cancer production. The New York Attorney General sued the United States government to stop the air shipments. He was indeed concerned after Dr. Marvin Resnikoff and I calculated that a crash causing dispersal of plutonium COULD KILL ALL 8,000,000 PEOPLE IN NEW YORK CITY.

Dr. Robert Barker, Health Physicist at the Nuclear Regulatory Commission, submitted an affidavit in rebuttal to Dr. Resnikoff's dispersal calculations. I quote this health physicist's affidavit:

> "The wind blows in a northwesterly direction, toward Manhattan, on the average *only about 3% of the time*." (Emphasis added).

He's giving odds of about 30 to 1 that such a crash would NOT kill everyone in New York City. Those odds might be amusing at the race track, but not with 8 million lives at stake. Isn't that elegant "protection" of public health offered by a health physicist?

Fortunately, Representative Scheuer introduced a bill into the United States Congress to stop this *particular* instance of the plutonium madness which is licensed by the Nuclear Regulatory Commission. Both houses of Congress passed the bill, and President Ford signed it into law. Occasionally, even Congress comes to life if the idiocy of a situation is sufficient.

And Some Nonsense About Radium

"But," say the nuclear proponents, "why all this *worry* about dispersing plutonium, when there is so much more *radium* per square mile of earth than the plutonium we expect to deposit?" To this nonsense, there are at least two sufficient answers:

(1.) It is the *airborne* plutonium settling to the ground which is the real inhalation hazard. What radium is six inches or twelve inches *below* the surface of the Earth is irrelevant, so long as it stays there. I certainly hope the nuclear industry has no plans to dig up the top foot of soil of the world, extract the radium, and spread it around on the surface in order to make plutonium a small problem compared with all that radium!

(2.) It is from the surface that the second hazard of plutonium-dispersal comes, namely re-suspension of plutonium by wind, vehicles, and pedestrians. Now, a three-inch boulder containing some radium is not very likely to become airborne. If it *does*, inhalation of the boulder just would be a bit difficult. So much for the ridiculous assertions about that top foot of the Earth, and its radium content.

Courtesy of Geoffrey Murphy and Minnesota PIRG *Statewatch*.

Embracing Our Radioactive Garbage

I won't bore you with the very long time that radioactive waste must be isolated from people. Everyone by now knows all about that. Instead, I prefer to regale you with some of the marvelous proposals of the nuclear energy advocates for dealing with this problem.

Since this stuff is dangerous enough to wipe out every man, woman, and child on earth many times over if it gets around, many have proposed careful burial of the waste. "But," say some of the nuclear advocates, "why bury it and guard it? This material is too valuable to bury. Instead we'll USE it in many, many ways in our economy, in our industries, and even in our homes." One early suggestion was to make plutonium-powered coffee pots. (Look, Mom, no electric cord for the percolator!)

There is certainly one marvelous solution for solving the radioactive waste problem—just USE it everywhere in the USA. I shall leave it to you to calculate the expectancy that 1% won't get out, with this superbly brilliant scheme.

Smoke-Detectors

I am appalled by the sale of smoke-detectors of the "ionization" type, which use radioactive substances inside, because an equally good job of saving lives from fire can be performed by the smoke-detectors of the photoelectric type.

The ionization-type detectors are not dangerous while they are in place—you could stand under them day and night, if you like, and they would not hurt you. The danger comes later, much later. For who is going to control the disposal and guarding of millions of such detectors, which will still be radioactive long, long after the houses have crumbled or been torn down?

The answer is: no one. When your house comes down—and eventually, it must—the radioactive substances will no longer be safely isolated on your ceilings—they'll be leaching out of a landfill in a rainfall, or wafting into the air as part of the smoke from a garbage-dump fire. Loose in the environment, the poisons will kill people.

Thanks to effective smoke-detectors which do *not* use radioactivity, we do *not* have to choose between saving our own lives today, or inflicting death on some of our descendants.

The ionization smoke detector is a cruel product, conceived by the irresponsible "let's love radioactivity" mentality. Naturally, the product received its license from the same crowd which bestows licenses on nuclear power plants. In both cases, the licenses should be revoked. And we're working on that!

Medical Experience for Engineers

In closing, let us examine why the nuclear proponents are so relaxed about the health hazards of radiation. Dr. E.F. Schumacher, the brilliant economist-philosopher, has said it isn't science and technology which any thinking

"First we have to convince the people that good health isn't everything.

person should oppose. Science and technology can indeed do marvelous things for us. But Dr. Schumacher has suggested he'd like to see "science and technology with a human face."

A million cancer deaths are, after all, just a number. Scientists deal with numbers all the time, so 1,000,000 isn't a disturbing number *per se*. In my earlier medical career, I used to work with cancer and leukemia patients extensively. I served as personal physician to some 30 or 40 of them in the last one to six months of their lives. It might help if every scientist and engineer had that opportunity as part of his (her) education. It is important to know what lives . . . and breathes . . . and dies behind a statistic.

Later, I spent two years doing studies on trace elements in a variety of mentally retarded children, at Sonoma State Hospital. I had a couple of days a week in the wards, seeing the human results of genetic damage. These children didn't look at all like statistics.

In 1969, after my colleague, Dr. Tamplin, and I had said that 32,000 extra cancer deaths per year in this country would be caused *if* people received the legally permitted dose of radiation from nuclear energy, we recommended that the radiation standards be made much tighter. Dr. Michael May, then Director of the Lawrence Livermore Laboratory where I worked, visited me in my office. Clearly he had experienced intense pressure from the AEC. In all my experiences with Dr. May, I had found him to be a fine person and a first-class scientist.

"Jack," he said, "I defend absolutely your right, in fact your duty, to calculate that a certain amount of radiation will cause 32,000 extra deaths per year from cancer."

But to my disappointment, he then asked: what makes you think that 32,000 would be *too many*? I must presume he was thinking in terms of the hoped-for benefits of nuclear power . . . technology without a human face.

"Mike," I said, "the reason is very simple. If I find myself thinking that 32,000 cancer deaths per year is NOT too many, I'll dust off my medical diploma, take it back to the Dean of the Medical School where I graduated, hand the diploma to the Dean and say, 'I don't deserve this diploma'."

□ □ □ □ □

2.

A SANE SOLUTION TO THE ENERGY PROBLEM

Some of you may wonder why I chose as a title for this talk, "A Sane Solution to the Energy Problem." Briefly, the answer is that I have no doubt whatever that the so-called "energy crisis" is going to be solved. No doubt at all. The only question is whether a sane, or an insane, solution is going to be sought.

It is clear to me that the President of the United States has already made his intentions clear to opt for an insane solution, just as have the last two Presidents before him. Now, since most people wouldn't be prepared to believe that all three recent Presidents of the United States are irrational—which would be the presumption if they choose an irrational solution to a major problem—we must next ask ourselves what we mean by the term "sanity."

It is when we examine the answer to that question, that we begin to understand the real basis for policy-formation by top governmental and private-sector leaders of our society. The key issue is "sanity" viewed through *whose eyes.* When I say that an insane solution to the energy problem is being opted for, I mean insane from the point of view of the overwhelming majority of humans, from the point of view of other living creatures, plant and animal, who are the hapless victims forced to co-exist on this planet with the human species, and from the point of view of the Earth as a planet capable of sustaining life over a long period of time.

There exists a very small minority of humans for whom the policies being chosen are quite rational, for whom the energy choices being made are indeed quite sane. The interests of this small minority are hardly consistent with those of ordinary mortals, but—if history is any guide—the interests of this small minority count for everything, at least as this minority sees the world. And because of certain features which characterize this minority, the diabolical situation is created in which the large majority is led down the path to its own disaster, and believes such a lemming-rush to the sea is simply marvelous.

The Name of the Game

Many people think that national objectives are to create a healthy economy, with full employment, with a healthful environment, and with as many people as possible enjoying their brief span of existence on the Earth. To aid in achieving this desired state of affairs, we have government dedicated to accomplishing these objectives.

That is a lovely view of the way things are—bearing as much resemblance to reality as Mr. Nixon's publicly-expressed view of his role in the Watergate cover-up. As I see the scene, there has always been a very different way in which the world really operates. With essentially no exceptions I am cognizant of, societies are operated by and for the interests of a privilege-elite. That is true whether the societies are called free enterprise, capitalist, socialist, communist, or by some other name. The operation of the society is still the same, namely the *maintenance* of privilege for its elite, and the *extension* of privilege for its elite.

The proper role of government has always been, is, and—unless a miracle supervenes—will always be, to insure that such privilege is maintained and *extended.* Assuredly that has been, and still is, the exact function of the United States Government.

Courtesy of Jules Feiffer.

If you doubt that, simply look at tax structures, look at the tremendous interest of the Congress and Administration in creating safe workplaces in industry, and at their enormous interest in maintaining environmental integrity. And by no means is that a description exclusively for the USA. I think the Soviet Politburo is just as much concerned about the welfare of the average Soviet citizen as is the United States Congress about your welfare.

As the privilege-elite sees a proper world, the peons should occupy their proper place, which is laboring hard at jobs they are generously provided, in a manner which guarantees that the elite shall maintain the largest share of the fruits of all labor, and that the share held by the elite shall be increased whenever and wherever possible. And the elite is quick to point out to the peons that this is assuredly the best of all possible worlds, for by the magic of TRICKLE-DOWN and a constant growth in the economy, the peons will enjoy an ever-better life than did their ancestors.

The belief that some individuals have a *right* to extra privileges is very, very widely accepted both in human society and other animal societies. It almost always requires *coercion* to enforce other notions, such as "from each according to his ability; to each according to his need", and similar equalizing policies. Instead, we accept the notion that those born smarter, healthier, or in an ecologically safer area, have a "right" to more wealth than people born less lucky. Some people say, "That's just the way the cookie crumbles", and others say, if they are among the lucky, "There, but for the grace of God, go I."

Whether or not human society should try to compensate its unlucky members for the inherently unfair distribution of talent, health, and other resources, is a moral issue of great importance and extreme murkiness.

All that is presently clear to me is that the privilege-elite has NO right to use coercion, physical force, or deception to increase its privileges. Yet throughout history, the privilege-elite on the whole has invariably used *all* of these unacceptable mechanisms, and continues to do so.

A major achievement of the elite—and further evidence of its intelligence—is making sure that the used do not recognize the user. Magic is a very old art, and the magician is supremely skillful at diverting attention from what is really happening. In society as a whole, the attention of the used is diverted again and again from the real users. The Ku Klux Klan and the Nazi Party are two extreme examples of misdirected focus and anger. Less sensational magic has been performed when voters blame the corruption of politicians on the high cost of election campaigns, when segments of the public blame inflation on unions, welfare recipients, or environmentalists, and when peaceniks blame the prospect of nuclear war on generals in the Pentagon.

Professionals as Apologists

In the early days of feudal societies, the elite could maintain its power by small private armies and tax-collectors, and the peons were generally kept in line by these agents, oftimes quite brutally. In the more modern, highly technologically-based societies, the armies are still there, the tax collectors are still there, but an important *new* class of agents for the elite has been created. This class is constituted of masters of business administration, scientists, academicians (and the universities they populate), and sundry professionals such as lawyers, physicians, and engineers. To this "professional" class, generously cut in for a modest share of the spoils, falls the task of convincing the larger society that everything is being done to assure a *good life* for that larger society. The pay has to be relatively high, because the job commonly requires the sacrifice of intellectual honesty.

On the other hand, we have all witnessed politicians, economists, and others—with obviously *sincere* concerns for humanity—who are nevertheless acting as team-members and automatic apologists for some anti-human systems. This is the sincere-sycophant syndrome. It is the natural consequence of an educational system which features the political philosophies of Plato, Machiavelli, Rousseau, Hamilton, Madison, and even Marx, but seldom if ever the libertarian classics of John Trenchard, Thomas Gordon, Lysander Spooner, and Albert Jay Nock. I, too, am a product of that educational system . . . and I have certainly made my share of mistakes.

This restricted education means that an adequate supply of truly innocent graduates is turned out annually who will not even *realize* it when they are being used later on. So widespread is the sincere-sycophant syndrome that one case per night is likely to pop out at you from some TV news-story.

The Patient-Doctor Relationship

So it is that Tony Mazzocchi (Oil, Chemical and Atomic Workers), one of the few true friends of the laboring man, was recently able to relate a story from Paul Brodeur's book, ''Expendable Americans'', about a *physician* who knew that a particular plant was killing its workers through asbestos exposure, but did absolutely nothing to inform the workers of what was going on. The passage from Mazzocchi's article is so choice that I'll quote it for you. Mazzocchi is speaking:

> When Dr. Grant [Medical Director of Pittsburgh Plate Glass and Corning] was questioned by Paul Brodeur—I know it's true because I was standing right there—Brodeur asked, ''Didn't you think you had a responsibility to tell the workers what was happening?''
>
> He [Dr. Grant] answered, ''To do that would be to violate the patient-doctor relationship.''
>
> And Brodeur said, ''Who's the patient?''
>
> Grant said, ''Pittsburgh Corning.''

Mazzocchi adds that *that* was one of Dr. Grant's more candid moments. So much for the ideals and behavior of some of my brethren in the medical profession.

Pesky Facts, Predetermined Conclusions

As for the engineers, the story is not much different. I'm sure all of you are aware of the embarrassment of the AEC over the existence of Wash-740 (the Brookhaven Report) and over the withheld *update* of the Brookhaven report, which indicated that casualties from a nuclear plant accident could be EVEN WORSE than those estimated in Wash-740. Conveniently, a group of unbiased engineers was assembled to produce the now-infamous Rasmussen Report, which through the application of ink to paper, suddenly made nuclear power the safest industry imaginable.

Never mind that the Union of Concerned Scientists procured documentary material which showed that an AEC staffer had written his boss an interesting memo expressing worry that "The facts may not support our predetermined conclusions." And later, this same man was writing that "The information we seek should not have the effect of raising unanswerable questions." Deborah Shapley of *Science* magazine interviewed Professor Rasmussen and others involved in the Reactor Safety Study. "They have maintained," writes Shapley, "that while the atmosphere surrounding the study may have been biased, the study itself was not."

Perhaps the most prominent hazard about the Rasmussen Report is the risk that people will believe it. Certainly the chance of its being right is less than the chance of being killed by a meteor. So much for objective, unbiased engineers.

[*Note:* On January 19, 1979, even the Nuclear Regulatory Commission repudiated its earlier endorsement of the Rasmussen Report. In 1975, the NRC called it "an objective and meaningful estimate of present risks associated with the operation of present-day light water reactors in the

Reprinted by permission of Chip Bok and the American Syndicate.

U.S." In 1979, the NRC admits the Rasmussen Report "greatly understated" the range of chances for a nuclear accident. The proper translation is what nuclear critics have been saying all along, before and after the Rasmussen Report: no one has the foggiest notion what the probability is of major nuclear power accidents. The NRC has also admitted that the Rasmussen Report staff conducted "inadequate" peer review of its methods and conclusions. Now we should take the advice of a great public servant who said, before Watergate, "Watch what we *do*, not what we *say*." Watch carefully for the NRC to deny or revoke the *licenses* for nuclear power plants—but don't watch with bated breath, and have smelling salts ready if it ever happens.]

Death As a No-No

For some strange reason, biologists have enjoyed an aura of respectability. *There* is a group of people one could trust! The AEC funded some twenty laboratories for twenty years, at a rate of 50 to 90 million dollars per year, to study the biological hazards of ionizing radiation. Death is an important biological hazard. Yet in that entire period, from that whole effort, there were probably fewer than 10 pages of studies which even mentioned the possibility of deaths from cancer from the peaceful uses of the atom. When Arthur Tamplin and I started to calculate deaths from radiation exposure (a task *we* interpreted as our assigned bio-medical mission), the entire AEC—particularly its prominent staff biologists—began to suffer from widespread internal hemorrhages. The garbage with which they endeavored to stem the flow of their bleeding, was a sight to behold. So much for how you can trust the biologists.

Why do I bring all this up here? It is primarily to make the point that the cadre of people who have comfortable sinecures in government, industry, and universities, are an important adjunct in modern society to the armies and tax-collectors in ensuring that the peons of society are kept in line. It should be obvious that I think moral cripples are to be found in AND outside the privilege-elite.

The function of sycophantic professionals is not suppression, but rather it is the donation of a *legitimate appearance* to those programs which the privilege-elite concocts to maintain and increase its privileges.

"Compensation" vs. Prevention

Thus the elite has no trouble finding lawyers who are eagerly "beavering" to establish a *right* to give people cancer with impunity! Here's what you have to watch out for:

It is medically impossible to prove that a *particular* cancer was caused by radiation, even when the radiation dose is *high*, since cancer can be induced by many agents. It *is* possible, however, to calculate the *probability* that a particular cancer was caused by radiation if you know how much radiation-dose was received.

When so-called "compensation" for radiation-injury is at issue, it is an outrage for any lawyer to argue that there should be *no* compensation for radiation-induced cancer simply because there can be no *certainty* that a particular cancer was radiation-induced. That argument simply amounts to giving private and government polluters an open-season on giving people cancer! Talk about a "permissive society"! We used to believe in severe penalties for killing people.

At the very least, a 43% probability that a person's particular cancer was radiation-induced, entitles the victim or his family to 43% of the so-called compensation, a 99% probability entitles the victim to 99% of the so-called compensation, etc.

The *bigger* moral issue, however, is *prevention* rather than punishment. And there may be a connection. Of course, we can count more lawyers who fight *both* prevention and punishment, than who fight for public health and the inalienable right of people to be free from physical abuse.

I must make sure there is no confusion about a related point.

It is certain that ionizing radiation can induce cancer in humans. That is proven, even at very low doses now, by comparing the incidence of cancer among people irradiated at one dose (e.g. only by natural radiation) with people irradiated at higher doses.

From such data, it is also possible to calculate the *number* of extra cancers which are occurring in a population as the result of exposure to various radiation doses (e.g., medical doses). It is just not possible to *name* the victims.

If *everyone* who is exposed to any radiation "got cancer", we would all be dead from natural radiation, of course. We all know that *not everyone* who is exposed to the influenza virus "gets the flu". Not everyone exposed to streptococcus gets a "strep" sore throat. Not everyone who has high levels of blood-cholesterol dies of a heart attack. Not everyone who skis breaks a leg. Not everyone who drives a car dies in an automobile crash.

Radiation-induced cancers are no different from other diseases or accidents: for an individual, they are a matter of statistical probability rather than certainty, while for a population as a whole, their occurrence is a *certainty*.

Act Against His Family's Interests?

During my experiences in the debate over nuclear energy and over alternatives to it, one question is very commonly asked of me when the actions of scientists, physicians, engineers, and physicists are impeached. "Why in the world would they DO that? It's their lives, their children's lives, which are at stake too. Surely these men, even opportunistic as they may be, would not knowingly jeopardize the lives of their families!"

That is an interesting point of view, and deserves very thoughtful consideration. It is this question which makes many people doubt that nuclear power *could* be as bad as the nuclear critics say it is. I've given a lot of thought to that question, and I can give you only an *empirical* answer. That answer is that such men must have an ability to make very effective rationalizations when they lie and otherwise obfuscate the facts they know.

A personal experience of my own might illuminate the ability to rationalize. Shortly after Tamplin and I stated that radiation would cause 20 times as many cancers per rem of exposure than the previous, most pessimistic estimates from the ICRP (International Commission on Radiological Protection), I received a phone call from an official of the U.S. Public Health Service, who requested that I never divulge his name. He stated that a high official of the AEC had come to see him, to ask that he help in the campaign "to destroy Gofman and Tamplin." He told the AEC official that he had read our paper and that, while he did not agree with everything in our calculations, in the main he thought our evaluation was quite reasonable. The AEC official, he said, pointed out that the issue was *not* whether Gofman and Tamplin were correct, but rather that the nuclear power program would suffer if their findings were not discredited. And, according to the Public Health official, the AEC man added, "Look, when all those cancers and leukemias start to appear, you'll be retired, I'll be retired, so what difference will it make? Right now, we need to protect the nuclear power program." The AEC official had children, a wife, and, for all I know, may even have had grandchildren. But none of that served as a deterrent.

This episode taught me never to trust people simply because they should not act against their apparent self-interest. *The vastly overwhelming self-interest is the current job*, with all its perquisites and privileges. Death of *their* children from cancer or leukemia is quite effectively rationalized away.

Dr. Alvin Weinberg, formerly head of Oak Ridge National Laboratory, told us how to make some of these rationalizations when he pointed out that it was premature to worry about the cancer hazard from deadly, long-lived plutonium. After all, said Dr. Weinberg, cancer research is very active, and maybe a cure for cancer will be found. Dr. Weinberg did not mention the *more* likely prospect that a *cure* for cancer may never be found—after all, that would destroy the rationalization. On the surface, Dr. Weinberg's rationalizations may even sound reasonable to someone with the largest part of the cerebral cortex missing.

Besides, cancer is not even the whole story. When a poison is cancer-causing (carcinogenic), there is a high probability that it is also mutation-causing (mutagenic). And vice versa. Ionizing radiation is definitely both carcinogenic and mutagenic.

It is well appreciated that genetic injury can cause death in subsequent generations, and that there is a genetic component in such killers as heart disease and diabetes. Less well appreciated are non-lethal afflictions which almost certainly have a genetic component too.

The casual dumping of persistent pollutants into the biosphere by the "advanced" nations may already have cursed our descendants worldwide with a miserable load of genetic degradation. So it is almost impossible to be polite when people try to justify nuclear power by bleating forth pure fantasy about cures for whatever will ail us.

De-rationalization, and Naming Names

This brings me to an important first point in the effort to achieve a sane solution to the energy problem. STRIP AWAY RATIONALIZATIONS WHENEVER YOU CAN. Generally they are very thin—just enough to allow the man to look at himself in the mirror when he shaves, and enable him to say, "I'm *not* the world's worst bastard." I mentioned above the half-witted suggestion for "coping" with plutonium toxicity. Its ethics can not withstand a challenge.

There is a common rationalization which can truly lead us down to the moral pits: "If I *don't* do it, somebody else *will*." The statement itself is often an accurate one. But so what? When enough people accept that rationalization, we not only *guarantee* that it will be an accurate statement, but we also guarantee that no moral sense can survive in us. For there is *no* act so vile that it can not be "justified" with that wonderful excuse.

And beyond stripping away the rationalizations, BE SURE TO NAME NAMES of those who make them. It is very hard for people to look at themselves and admit they are rascals or nincompoops. While they may become even more defensive to protect their rationalizations, they will in this posture become even more absurd (assuming that were possible) and manage to discredit themselves *entirely*.

One of the favorite rationalizations of the scientists who serve as propaganda-vehicles for the nuclear industry, is that we should not worry about the long, radioactive half-life of plutonium-239—whose half-life is 24,400 years. With the pride of ignorance, these nuclear apologists happily point out that society handles tons and tons of lead, and the half-life of lead is *infinite*. That's meant to be a brilliant put-down of the people who worry about handling tons and tons of plutonium. Leonard Sagan, M.D., was perhaps the first to use this line widely.

At the same time that Dr. Sagan was saying this (1970), the *New England Journal of Medicine* carried an editorial stating that it was a national disgrace that 200,000 children in the USA carried overtly toxic levels of lead in their blood-streams, and that these children were suffering from permanent, irreversible, lead-induced BRAIN DAMAGE. More recently, an article in *Pediatrics Digest* (1977) points out that the number of children suffering from lead intoxication in the USA is more like 600,000 rather than 200,000. And these dismal figures exclude all the American *adults* who carry sub-clinical levels of lead in their blood, from auto exhausts and industry. These lead-levels may well account for the fact that about one-half our population is as nutty as it is.

The inordinately stupid and irresponsible handling of lead by our society is undoubtedly one of the worst of societal crimes, with the greatest injury and hence injustice concentrating in the ghetto, on the poor.

For nuclear apologists to justify the stupidity of handling tons of plutonium (which is what a major nuclear power program requires), by stating that we handle tons of lead, is simply too irresponsible to withstand a challenge. *That* rationalization can be publicly destroyed, and so can the *other* fragile fabrications of the well-paid cadre of sycophants who try to justify the depredations of the privilege-elite.

The Charm of NUCLEAR Electricity

Thus far, we have focused on the use of the technical community and of certain other professionals by the privilege-elite. Now we must turn to the question of *why* the privilege-elite seems so determined to push an anti-societal technology upon the public. There is no *a priori* reason that an insane technology such as nuclear power should be chosen by the privilege-elite. Indeed, it would not be unthinkable, all other things being equal, that Mr. Carter and Dr. Schlesinger might choose to solve the energy problem by *sensible* means. But they don't do so, and there must be a compelling reason for the choice of insanity.

Recall, I said earlier that one must always ask, "Sane or insane *for whom*?" Even if we were to shrug off the fact that "peaceful" nuclear power means a world laced with atom-bombs, there are other reasons why nuclear power can never be safe with respect to the great bulk of people. Consider HEALTH.

Radioactive Containment . . . by Robots?

If all radioactivity from nuclear power were perfectly contained, under all circumstances (normal and abnormal), then it would be hard to say that nuclear power is an insane choice from the viewpoint of public health. For if every bit of radioactivity were perfectly contained, no one would get hurt, and nuclear power would be "safe," even if inordinately expensive.

Now, even the nuclear advocates don't claim they can contain the radioactivity perfectly. They only claim containment of such substances as plutonium to a degree of perfection of one part in ten million, or one part in one billion. Not absolutely perfect. Even a low moron would consider that sort of claim to emanate from people far less brilliant than he.

But it is the *need* for that sort of claim which makes nuclear power an *absurd* and *insane* technology. The required containment has not been achieved, and there is no prospect it is going to be achieved.

I should back off and correct myself a bit. Dr. Alvin Weinberg has outlined the conditions under which

'Listen, We Could Get Burned'

Engelhardt in the *St. Louis Post-Dispatch*.

nuclear power might be safe, namely with absolute and perfect social tranquillity (*Science,* July 7, 1972). A totally robotized society would indeed help eliminate the human error and malevolence which could interfere with perfect containment of radioactivity. But even this type of ideal society (ideal in the Weinberg sense) would still be plagued with machine failures and Acts of God.

And one might surmise that there is something left unsaid in the Weinberg dream of a nuclear economy: Robotized humans are not likely to complain about dying prematurely of radiation-induced cancer or leukemia in any case. So that is a plus for Weinberg's ideal "socially tranquil" society.

How does the privilege-elite view the question? The answer is that a robotized and servile society is exactly what would suit it best!

From Control to Tribute

In the early, heady days of a free enterprise economy, there was growth, innovation, inventiveness, risk-taking. Those who were the best entrepreneurs did exceedingly well in acquiring privilege, material wealth, and once having acquired a fair share of both, they could go on to extend their acquisitions. But much of that heady phase is over, and the going is tougher. It is not that the privilege-elite no longer aspires to the ownership of everything. It is just that it doesn't want to take risks. It much prefers a sure thing.

Besides that, the privilege elite feels it has already given too much of the wealth to labor and to its sycophants; the name of the game is now to get as much *back* as possible. For these purposes, we now find the massive conglomerates of corporate power endeavoring to bring one segment after another of the economy *under their centralized control*, so that they can insure themselves a guaranteed profit, at the expense of the larger public, from each segment of the economy. We don't see new entrepreneurs entering the automobile business each year. We find the food growing-and-distributing functions becoming centralized into fewer and fewer hands. There aren't many new telephone companies being organized each year. And as for energy—be it solid fuels, liquid fuels, or electricity—this is not the arena for the wild-catter of old. Of and by *itself*, bigness is not necessarily bad. There can well be certain economies of scale which might be beneficial to the larger public. It is not the bigness which is at issue. It is, rather, the setting up of our economy in a manner such that, at every turn, the privilege-elite has an absolute guarantee that all members of the public PAY THEIR TITHES. And the tribute-payment is steep, and getting steeper with every passing day. It is paid in many, many ways.

Unemployment . . . Too Bad!

When the privilege-elite discovered that a bigger profit could be obtained by replacing labor with energy, it went right ahead to create unemployment with the shiboleth that new growth would create new jobs. Yet today, with all the growth which has occurred, we have the largest, most protracted unemployment since the Great Depression. They've even coined a phrase for it, "structural unemployment," as though it were ordained by the Almighty.

Note well who suffers the most from this structural unemployment, whose hopes and aspirations are shattered with the destiny of staying in the ghetto without the hope of leaving it. And note especially well *who pays* for this social disgrace. It is *not* the large corporations, nor the privilige-elite who owns them. It is Mr. and Mrs. John Q. Public, through a tax burden which requires them to run faster and faster to get back to the starting line. All this, of course, is arranged through the brilliant intercession of the Congress and the Executive Branch, under such programs as "welfare" and "unemployment compensation."

"Partnership" with the Public

Not satisfied with having the segment of the public which *does* work, pay for those who *can't* find work, the privilege-elite has extracted a number of further guarantees that its privilege will be extended. It now demands that the research and development for the new products be financed also by John Q. Public, through his tax-dollars. So we find that when a new energy system such as the breeder is proposed, we hear that it will be "commercialized" through the joint efforts of government and private industry. A beautiful balance is struck. The entire electric industry will pay 250 million dollars, and the Federal Government—meaning you and me—will pay 9 billion, 750 million dollars for this "commercialization" effort. That's real equity, real private enterprise; that's what makes the American economy grow and prosper.

So, not only does the privilege-elite steadily tighten the noose of absolute control over virtually every major part of the public's expenditure-dollar, it even wants the public to pay the bill to explore any of the potential changes in the mix of products on which the tithe will be exacted. AND to pay the bill for the expendable part of the public which is left out entirely! Food, transportation, and energy are big chunks of the average person's expenditures, and as the privilege-elite sees the ideal world, it wants to insure a guaranteed return from each of these expenditure-areas. And it wants nothing left to chance in this endeavor. It wants to control the source of supply, and to price everything right up to a point just *short* of outright revolt by a robotized public.

The Trouble with Solar Energy

If one were trying to think of how to handle the energy question *with the goals just outlined*, one could be sure the choice would *not* be decentralized solar energy for heating, cooling, or electricity! *That* choice would make sense from the point of view of the public peon, but it certainly is pure insanity from the point of view of the privilege-elite. From the point of view of the privilege-elite, the only choice which makes sense is to opt for the energy approach which absolutely guarantees complete dependence on a central source. *That is the explanation for the miraculous romance between our energy-planners and electricity.*

Never mind that there is no reason which makes *sense* for using electricity for more than about 10 % of our end-uses of energy. Electricity makes one hell of a lot of sense if you are trying to guarantee the public's *dependence*, if you are trying to guarantee that everyone will pay his tithe with the energy part of his dollars.

So it is no accident that the ERDA or the Department of Energy—which is a very, very large sycophant organization which works for the privilege-elite—spends by far the largest share of its budget on central-station power, on trying to convince everyone that *using* electricity, where electricity makes no sense on a thermodynamic basis, is simply the most marvelous idea. And even when the ERDA dabbles in solar power, most of its support goes for *centralized* solar electric systems!

Nukes As the Final Solution

If you go one step further, after deciding that doing just about everything *electrically* is an excellent guarantee that everyone pays his tithe, the next step is to ensure that electricity-generation has to be forever centralized SECURELY. Nuclear power came along "in the nick of time" (to borrow nuclear salesman Glenn Seaborg's catchiest line) to solve that problem. Nuclear power is just what the doctor ordered. It provides guaranteed centralized control; it won't *ever* be a do-it-yourself technology; and every dollar poured into the super-costly nuclear electricity option helps foreclose the development of challenges from decentralized energy systems. It is indeed ideal! And that is why the ERDA is so pleased to spend just about everything it *can* on this technology. What better way can ERDA's devoted boot-lickers hope to curry the favor of their masters?

And once having tasted the possibility that a final solution has been reached for guaranteeing a stable tithe from everyone on the energy part of his or her daily expenditures, the privilege-elite can be expected to fight tooth and nail against LOSS of this handy payment. As for hazards to public health, I've already covered the manifold rationalizations which are used. Technology will solve all the unsolvable problems. As for the privilege-elite itself, or its families, suffering from side effects, this threat is of small moment when viewed alongside the stacks of dollars to be had.

\+ + + + + + + + + + + + + + +

(CWIP = Construction Work in Progress = That part of your electric bill which finances the construction of nukes = The conversion of all customers into involuntary investors who receive no return on their investment.)

\+ + + + + + + + + + + + + + +

From *Mel-Practice in New Hampshire; A Cartoonist's View of Governor Meldrim Thomson*, by D.B. Johnson. Meredith, N.H.: Intervale Publishing, 1978. Reprinted by permission of its author.

So it is that an insane technology, from your point of view or mine, is the darling choice of the privilege elite. That the Congress and the Executive Branch of government lend their full support to this insanity is not at all surprising, once one is clear in his head that these branches of government have *nothing to do with* serving the best interests of the larger public. They never *have* had, and they don't *now.*

Jimmy's Phony Promise

Every once in a while, when the peons appear restive over the massive shellacking they are taking with inflation and unemployment-payments and taxation, it becomes the right time for a little infusion of populism. This is particularly helpful once every four years. The populism is required just up to election day. Then it all disappears.

The most recent illustration, where a little populism was required, was the Jimmy Carter campaign when he sensed that the peons were a bit restive about the hazards of becoming victims of nuclear power. Nuclear power will be used only "as a last resort," Mr. Carter assured. With the election safely passed, the need for populism disappeared. Some foolish person had the temerity to ask Mr. Carter's energy staffers about the broken promise of treating nuclear power "as a last resort." Unabashedly the staffers provided the answer: "Of course nuclear power is a last resort . . . and we are down to last resorts!"

So, "Anchors Aweigh." Some people seem to feel a little deceived. But that is only due to their naivete about the real function of politicians, and of government in general. If the privilege-elite wants *nuclear* electricity as a way of guaranteeing a tithe from everyone, you can count on politicians—Republican, Democrat, Martian, or otherwise—to bend every effort to see that the privilege-elite is accommodated. *

In the real world where the movers and shakers make things happen, there is no energy crisis worth talking about. The only crisis worth talking about, and doing something about, is ensuring that everyone pays the energy-

\+ + + + + + + + + + + + + + +

* Generalizations have exceptions, of course. To his everlasting credit, U.S. Senator Mike Gravel became the Congress's leading and unwavering opponent of nuclear power, long before it was "reputable" to do so. And in all states, the laws of statistical probability ensure that there will always be *some* politicians who do not fit the generalization. But when you want to know "where the action is", watch what the 98 are up to, and not what the TWO are doing!

tithe forever, just as they are squeezed to pay the tithe for other essentials of life. So there is going to be no move toward sanity from the movers and shakers. "And you can depend on that."

Sanity for the Public: How?

Some have hoped that the education of the public about the hazards of this absurd technology would lead to its rejection. Some even thought education of Congress would lead Congress to reject the technology by law. I must admit that even I *once* thought there was some virtue in trying to educate the Congress. I think we all know by now what value there is in *that* endeavor. As for education of the public, that is infinitely more valuable than educating the Congress, simply because the two *march to different drummers*.

By now, we have succeeded in educating the U.S. public to the hazards of nuclear energy. The polls show that the public is convinced that nuclear power is the riskiest choice of all the energy options. BUT the same polls show that about 65% of the public is convinced that we must go ahead with nuclear power, because otherwise the economy will falter, jobs will all be lost as the economy goes to hell, and we will have to return to the caves!

The public is no more immune to the threat of economic punishment than the sycophants who are well-paid for their obeisance to the masters. Survival is indeed important to people, and anyone who thinks people are going to jeopardize their survival is simply mistaken. So, if a slick Madison Avenue campaign has convinced people that the economy needs nuclear power, it is going to be very hard to convince them otherwise. To be sure, the peons do become restive the more they realize what a rip-off nuclear power is, as they pay their utility bills. But unless they can see a way out, even this restiveness doesn't help very much.

"Civil Disobedience"

There are those who feel that a dramatization of the issue by direct action—for instance, "occupying" nuclear plant sites or blocking their entry gates—will block further development of nuclear power. There is certainly no doubt that such dramatizations do indeed have educational value, and they do help raise the consciousness of a segment of the public about the massive rip-off of their *lives* which centralized control of society imposes. But, at the same time, a large segment of the public will see such direct action as a threat to livelihoods and the economy, and some will interpret it as unfair to business; and these segments will applaud repressive measures to counter the threat. Then the centralized control of our lives will not only be via the guaranteed economic rip-off, but there will be even more invasion of our residual rights as humans than already exists, through dossiers, wiretaps, infiltrators into the environmental movement, blacklists for employment or grants, and the many other techniques for creating a Thousand Year Reich.

It is hard to see the way back to a free society, to real private *enterprise*, and to some restoration of the chance to exert control over our own lives. These goals are totally at variance with the goals of the privilege-elite, and its members have the power and financial means to impose their goals—no matter that they make the earth unfit for human or other habitation. In his book, *The Greening of America*, Charles Reich saw the development of Consciousness III at a rate a bit higher than has turned out to be the case.

Indignation over Dying

But there are some possibilities which I think are constructive, and the energy problem may be the issue which helps move society *away* from the total robotization currently planned for it.

First of all, there is already occurring in the chemical industry the rash of cancers which was quite predictable from the helter-skelter introduction of chemical carcinogens into the workplace and into the environment in general. There will soon be a rash of cancers and leukemias in the nuclear industry. It will be important to try to get people to appreciate the rip-off of their health in these occurrences. As yet, labor has accepted the cancers as the price of employment, but it is possible that workers may still become indignant about such a royal screwing. Especially important will be to identify *by name* through whose courtesy such marvelous, premature deaths are brought to labor. The same holds for identification by name of the promoters of nuclear power, who should be brought to account for the cancers which are going to ravage the workers of the nuclear industry.

By Everdell. By permission of Community Press Features, Boston 02116.

It is sad to contemplate that we will probably have to contaminate the planet irreversibly by radioactivity to some degree, before much of the public can be brought to indignation.

On the positive side is the fact that nuclear power is very expensive now, and getting more expensive every day. There is still some degree of entrepreneurial activity in the country, not under monopoly control, and there is still very great inventiveness in the population. It is conceivable that the combination of inventiveness and entrepreneurism may show that energy-efficiency and solar energy can be made cheap enough, and reliable enough, that the restive peons may embrace them to the dismay of the Schlesingers and the Carters who support total control of our lives.

Courtesy of Tom Chalkley Graphics, Washington, D.C.

Therefore, it is very important to support the efforts by amateurs, by small companies, and by inventors, to solve our energy problem *by going around the central control systems*. And it is important in doing this, rigorously to avoid claiming success for the systems which are not soundly tested, and vigorously to help people avoid investing in systems which don't really work. Getting a bad name from *poor* decentralized systems would be moving backward. But I believe that small, practical, individualized energy-systems are coming along, and that they may show up the ludicrous nature of the federal government's energy priorities.

As I would see the effort to restore our rights to choice, to restore our right to live without being robots for someone's privilege-system—in fact, to restore our dignity as free humans—there are two efforts which are going to be needed. One is at the local level, and the other is at the national level.

At the local level, small-time functionaries will do everything they can to block any effort to help us out from the tithe-system and from control over our lives. This is only natural, since small-time functionaries see themselves as faithful sycophants who aspire to become big-time functionaries—sharing in such plush grab-bags as the Koreagate pie. But while they are still small-time functionaries, they have to live in your midst, and they are, therefore, vulnerable to the constant exposure you give them, as they serve against the public interest.

For example, as local legislators and public utility commissioners attempt to block the introduction of sane energy strategies for individual users by zoning ordinances, building codes, and the like, they can be exposed, and *named*, and held up for public ridicule. I think this is very important and can be effective.

If this approach is used, even their families will come to wonder what makes them such SOBs. There is already, sporadically but surely, evidence that local officials can be brought to act *for* the public interest, rather than against it. This is true even for some utility commissioners.

At the national level, the only constructive thing I can think of is for everyone to realize what a thwart the federal government represents to the health and happiness of the population. To expect that the national government will work toward energy sanity, is empty folly.

That simply won't happen. If it helps your adrenaline level to write to your Congressman, by all means do so, but don't expect constructive action from him. Letters to Congressmen can help on issues concerning which the privilege-elite is totally *indifferent*. But let the issue touch the elite's interests, and then you'll find your Congressman can't hear you or read. I know of the campaigns to dis-elect "the dirty dozen," and I certainly have hoped that their dis-election would improve matters. But in the main, I think we get another dirty dozen with different labels and stripes.

The best thing we could achieve at the national level is a steady reduction in the size and power of the government over our lives. If enough people would consider decreasing the *size* of the federal government itself as a major national goal, there might be some real hope.

Just imagine the salutary effect of reducing DOE (the Department of Energy, formerly ERDA) and the NRC (the Nuclear Regulatory Commission) to half their current size. The obstruction they can render would at least be reduced in *half*, and because the dollars saved could stay at home where something *constructive* could be done with them, the chance of sane energy solutions would be materially enhanced—at least by a factor of two.

If DOE disappeared altogether, and the dollars could remain with you for energy progress, we probably could *guarantee* a rational solution of the energy problem.

Peg Averill, in *WIN Magazine* 1977.

□ □ □ □ □

3.

ON THE WAY TO THE BANK:

Or Why There Will Never Be a Solution to the Radioactive Waste Problem

Recently President Carter, who got environmentalists' votes by making campaign promises to develop nuclear energy only as "a last resort", has embraced nuclear power with both arms, has appointed an avid nuclear promoter (and former chairman of the Atomic Energy Commission, 1971-1973) as Secretary of Energy, has called for an acceleration in nuclear power plant licensing, has announced he favors a nuclear breeder program, and has allocated enormous outlays for breeder research in his budget. *

Since essentially nothing has changed in the energy picture in the few months between the campaign promises and the energy policy announcements, one can only conclude that Jimmy Carter was planning a blueprint for a nuclear future all along.

And if any Jews are under the delusion that a gung-ho nuclear power program can free us from Arab-oil blackmail during the next 30 years, they had better bone up on some facts about the nuclear fuel shortage, before they

\+ + + + + + + + + + + + + + +

* The Carter Administration even favors *ramming* the radioactive waste down the throats of unwilling states, if that is what is required to keep the nuclear power program running. On January 25, 1979, the Department of Energy testified to Congress that it opposes state veto-power over federally selected radioactive burial-sites.

get used, too. Although the President talks as if additional nuclear plants could make a significant contribution to our energy supply during the next 30 years, that is simply not true. Due to our nuclear fuel shortage, the maximum contribution from building additional nuclear plants would be a trivial 4% of our gross energy supply (1).

Things Are Seldom What They Seem

Perhaps many of you are confused about the apparent inconsistency of Carter's *enthusiasm* for nuclear power which I have described, and his apparent *opposition* to the breeder project in Tennessee and to plutonium recycling in current nuclear power plants. The President generally claims that these latter positions are a reflection of his concern over the proliferation of nuclear weapons. However, so many *other* aspects of the President's nuclear program are clearly *proliferation stimulators*, that one cannot take very seriously an alleged concern over this issue in the Carter program.

The probable explanation for the President's opposition to the breeder planned for Clinch River, Tennessee, is that its particular design makes it an obsolete white elephant and an economic crock; both he and his Secretary of Energy have said so. To use Jim Schlesinger's assessment: All that the elaborate multibillion-dollar Breeder Project could demonstrate is, if one can produce electricity, it can be sold!

As for the rest of the President's nuclear energy plans, I deplore every aspect of them. Although they will prove highly *counter*-productive in helping to solve any aspects of the energy problem, they do help solve some problems for the ailing nuclear energy *companies.* There are many reasons for faulting the entire nuclear program of the

+ + + + + + + + + + + + + + +

(1) Gofman, J.W., "Gross Energy Available through Light Water Reactors", *Committee for Nuclear Responsibility Report 1977-2*, May 1977.

President (2).

A Counter-Productive Plan

First, we don't really need more central electric power plants for the health of our economy. We are already using *electricity* for purposes which would be far better and more economically served by *non*-electrical energy. But there is a set of power-pushers known (by environmentalists) as the Fools-for-Joules crowd.

Second, even the President has acknowledged that co-generation of electricity by industry could provide any additional electricity needed, far more cheaply than power provided by nuclear power plants. Co-generation would free up capital required for other segments of the economy, where it is really needed to create jobs and maintain the health of the economy. *

Third, the rapidly escalating costs of nuclear power have made it an economic disaster. This situation is so bad that the respected investment specialist, Saunders Miller, has stated: "From an economic standpoint alone, to rely upon nuclear fission as the primary source of our stationary energy supplies will constitute economic lunacy on a scale unparalleled in recorded history, and may lead to the economic Waterloo of the United States" (3). And Miller was not even discussing the fact that policy-makers are proposing more "nukes" than the number which would be *fuelable* for

\+ + + + + + + + + + + + + +

(2) Gofman, J.W., and O'Connor, Egan, "Jimmy Carter's Energy Plan: Myths vs. Realities" (Part I: Solar Energy; Part II:Energy Conservation; Part III: Nuclear Fission), *Committee for Nuclear Responsibility Reports 1977-3, 4, 5,* June 1977.

* The energy-potential of co-generation of power (generating electricity where industrial steam is going to be generated anyway) was estimated in May, 1978, as equivalent in the USA to 208 giant nuclear power plants. See *Industrial Co-generation*, by Robert Williams, Ph.D., Center for Environmental Studies (Report #66), Princeton University.

(3) Miller, Saunders, assisted by Craig Severance, *The Economics of Nuclear and Coal Power.* New York: Praeger Publishers, Inc., 1977.

their lifespans! Unless, of course, they plan to make us dependent also on foreign sources of uranium . . . .

Fourth, the President's program will not inhibit nuclear weapons proliferation; more likely it will accelerate such proliferation.

Fifth, the President talks about the safety of our current light-water reactor plants as though this safety were established, when in fact it simply is not. Far from it. Catastrophic accidents—like the melting of nuclear fuel-cores or the rupture of reactor vessels—are considered likely enough that the nuclear industry demands and obtains an act of Congress (the Price-Anderson Act) to limit its liability for such incidents. "Incidents" is Nuclear-Speak for accidents which could contaminate an area up to the size of California with radioactivity.

Core of the Controversy

I could go into any one of these issues in detail here, but I shall not—for one very specific reason. That reason stands out above all others, in my opinion, and if more people understood it, the whole character of the so-called nuclear power controversy would change:

> Nuclear power is simply unacceptable for this or for any other society, and there is no credible way that it can ever be *made* acceptable. I shall state that this conclusion is correct *even* if we were never to have a major nuclear power plant accident.

Other speakers will come and disagree with me. The public is often heard to comment that it is difficult to know what the truth is, because the experts disagree!

When two experts reach different conclusions from the same evidence, there are only two likely explanations. Either, one of them is using faulty reasoning (for instance, an inappropriate, illogical use of a statistical principle), or the two experts have chosen different assumptions to fill certain *gaps* in the existing evidence. The latter is

"Why couldn't you have just said you left your apple on the hot plate, instead of 'There's been a core meltdown'?"

Jim Berry.
Reprinted by permission of the
Newspaper Enterprise Association.

the case in the nuclear power controversy. When experts choose different assumptions, naturally they reach different conclusions. A scientist who strives for public enlightenment instead of public confusion, will explicitly clarify how his assumptions differ from those used by others.

I intend to show you that the whole nuclear power debate revolves around the assumptions made on just one point, and I shall clearly state the assumption I use and *why* I think it is the correct assumption.

Undoubtedly expert opinion has a necessary place in certain controversies. It may surprise you to hear me say that I think it has no necessary role in the nuclear power controversy. One need neither *be* an expert nor *listen* to an expert to arrive at the conclusion that nuclear power is unacceptable and will remain unacceptable. All that is required is some simple arithmetic and some common sense in choosing that crucial assumption I referred to. So let me suggest perhaps a new framework for your thinking about nuclear power—particularly about the problem of radioactive waste.

Monstrous Amounts of Poison

The operation of nuclear reactors generates astronomical quantities of radioactive garbage of several types, the amount of radioactivity generated being in direct proportion to the amount of electricity produced. In one year of operation, a 1000-megawatt nuclear power plant generates fission products (like strontium-90 and cesium-137) in a quantity equal to what is produced by the explosion of 23 megatons of nuclear fission bombs—or more than one thousand bombs of the Hiroshima-size*. And if people like Jimmy Carter and James Schlesinger can prevail, we will have 300 or 400 such plants in the USA alone by the year 2000.

+ + + + + + + + + + + + + + +

* Gofman, J.W., "The Fission-Product Equivalence between Nuclear Reactors and Nuclear Weapons", in *The Congressional Record* (Senate section), July 8, 1971. This is also a C.N.R. Report.

This means that *annually* we will generate the strontium-90 and cesium-137 garbage equivalent to a full-scale nuclear war, and we will go on doing this year after year until the fuel runs out. Should society experience the ultimate disaster of letting nuclear *breeders* generate unnecessary electric power, we would have the 1,000 or 2,000 plants fondly dreamed of by nuclear unrealists, because breeders solve the nuclear fuel shortage.

Whether we consider 300 or 2,000 nuclear power plants, the quantity of nuclear fission products to be produced makes the quantity produced in the entire *military* fission program miniscule in comparison, and these wastes are a serious enough problem of themselves.

There is *no* nuclear fuel, *no* reactor type, *no* operating mode which can significantly alter the astronomical quantities of radioactive nuclear fission products produced in nuclear power installations. The quantity is fixed by the immutable laws of physics. This is *not* a point of dispute among experts.

How much of the garbage can we afford to let loose into the environment? We have very good estimates of the numbers of cancers, leukemias, and genetic diseases which will occur per unit of radiation *exposure*. These "dose-effect" estimates are based on real-world observations plus certain assumptions; as the data-base has *increased*, the disagreement between experts on the toxicity of radiation has *decreased*(4). Even if I were to use the nuclear *advocates'*

+ + + + + + + + + + + + + + + +

(4) Gofman, J.W., and Tamplin, Arthur R., "Epidemiologic Studies of Carcinogenesis by Ionizing Radiation", *Proceedings of the Sixth Berkeley Symposium on Mathematical Statistics and Probability.* University of California Press, 1971.

Gofman, J.W., "Statement in the Matter of 10 CFR 51, Licensing of Production and Utilization Facilities; Environmental Effects of the Uranium Fuel Cycle". Submitted to the Nuclear Regulatory Commission for Docket # RM 50-3, by the Sierra Club (Buffalo, New York), October 3, 1977. This is also a C.N.R. Report.

Gofman, J.W., "The Question of Radiation Causation of Cancer in Hanford Workers", *Health Physics,* in press.

present estimates of the number of deaths per unit of radiation exposure, I would STILL consider nuclear power to be unacceptable. In other words, the harmfulness of radiation is pretty universally acknowledged now, and is not where the *crucial* assumptions are made in the nuclear debate.

The Crucial Assumption

The ultimate size of the epidemic of cancers or other diseases which nuclear power can cause, depends upon the size of the radiation *dose* which will be received by the public from nuclear pollution in the course of a nuclear energy economy. And of course, that radiation dose depends on the fraction of the radioactive inventory which will *escape* into the environment in which people try to live. This reasoning is accepted by all experts too.

It is the estimate of the *fraction escaping* which is the crucial assumption, and which is the core of the controversy. If the fraction of the radioactivity which escapes containment is ZERO, then there is simply nothing at all to worry about with respect to public health consequences; the cancers, the leukemias, and the genetic diseases would simply not occur.

But the dream-world in which the fraction released is *zero*, exists only in the imagination of nuclear promoters inside and outside the government. All other people have to live in the real world. Thus, nuclear energy represents history's grandest experiment in containment. The experimental subjects are all the humans on Earth. The booby prize for failure in perfect containment of the radioactivity, is a public health disaster which will make even current cancer fatalities appear trivial. And the grand prize for success is a little energy which can be better obtained by *other* means at *lower* cost, with ecological sanity, by means which are available to us if we choose to use them. But the President and his Secretary of Energy tell us to go-for-broke in this game of roulette.

The Problem's 4 Elements

To summarize so far: There are four elements to the problem:

1. How much radioactive poison is produced by nuclear power plants; no disagreement.
2. How toxic the radioactive poisons are; negligible disagreement now.
3. How much of the radioactive poison is going to *get* to people by escaping into the environment during all steps of the nuclear fuel-cycle, from mining virgin ore to handling the *used* fuel; enormous disagreement, though seldom discussed.
4. How much of the radioactive poison is going to get to people by escaping from some hypothetical "permanent" storage place *before* it has decayed to harmlessness; isolation from the biosphere is required for periods ranging from 600 years to 250,000 years and longer, depending upon which particular mixture of radionuclides is under discussion. Lots of discussion of this topic, but little agreement.

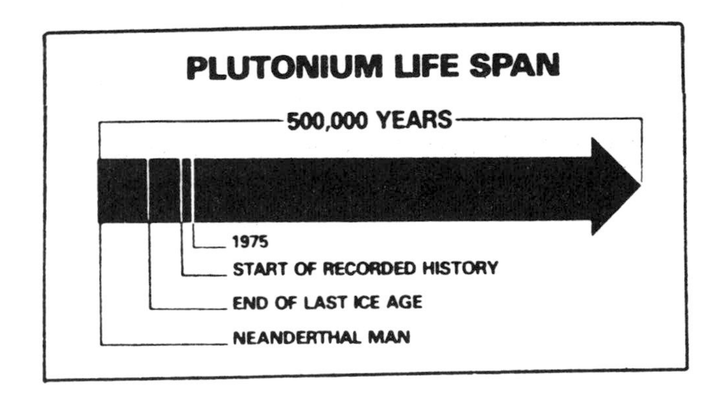

As we shall see, the *real* solution to the problem of radioactive poisons lies in the first element of the summary—namely *stopping* the production of the poisons, for the simple reason that there is no credible way in which the third element of the problem *will ever be solved*.

Engelhardt in the *St. Louis Post-Dispatch.*

'Quick, Look Over There—See What A Clean, Beautiful Plant We Have'

Not unexpectedly, the proponents of nuclear energy, including the industry and the U.S. Government, virtually never address the third element because they know they cannot produce credible answers to it. Instead, they want us to assume that by some miracle outside of human experience, the third element will solve *itself*, while they go on to discuss their efforts to solve the fourth problem (namely the so-called "permanent" waste storage).

Thus, the public is offered long and learned discussions about just where the radioactive waste will be buried, what form it will be in—such as liquid vs. solid—and whether it can be contained in the burial site for the requisite number of thousands of years.

This is *not* the primary problem. The *real* problem, for which there is not and never will be a credible solution, is what I choose to call the loss of radioactivity "on the way to the bank". And I shall discuss with you how very close to *zero* the losses on the way to the bank must be, in order to avoid a public health disaster. But first I will discuss the fine job of *diverting* public attention from recognizing the real problem.

Deliberate Diversions

It is no accident that the discussion promoted by the nuclear advocates carefully circumvents the most serious problem, namely how much radioactive garbage is lost *on the way* to the bank. The reason is simple enough. The public has demonstrated that it is extremely concerned about the generation of radioactive garbage by the nuclear industry. Even some government officials have expressed concern, which is itself something of a first-order miracle.

The public shows good sense in being concerned about this radioactive waste, and is not easily conned into believing hogwash on this issue. Therefore, from the promotional point of view, it is crucial that the public's concern be focused on the final resting place for the

'That's a good question which demands some real evasion.'

radioactive waste without ever raising, for public consideration, the realities about how much gets lost *on the way* to that final resting place, which I refer to as "the bank".

If the public gets to understand that the *unsolvable* problem is the extent of loss into the environment *on the way* to the resting place, the public will be horrified and likely demand the end of this pernicious technology. But if no one tells the public about the really hopeless part of the problem, the public's attention can be diverted to debates on the relative merits of burial in salt, burial in bedrock, or burial at sea. So successful is this ruse that even parts of the so-called anti-nuclear movement are taken in.

Some discussions of safe storage have ranged from the absurd to the utterly ridiculous.

Shipping It Out

We have Dr. James Schlesinger's proposal that we rocket the radioactive waste to the sun, where it simply can't get into our biosphere. This proposal was a classic illustration of ignoring the third element of the problem, in this case, how much of the radioactive garbage is lost *before* reaching that ultimate repository, the sun. Dr. Schlesinger just didn't reckon with occasional failures of garbage rockets. However, Battelle Pacific Northwest Labs apparently *did*, and commented wryly in 1974: "There seems to be a degree of uncertainty as to whether . . . a trajectory can be achieved with sufficient accuracy to insure that a capsule will not come back to Earth in an unplanned manner."

Something for Everyone

Another proposal, at the acme of brilliance and responsibility also, originated with Dr. Dixy Lee Ray, once chairperson of the AEC too (1973-1975). She suggested that we re-define the radioactive garbage as a *resource*, rather than as waste (5). Having suggested this solution-by-definition, she proposed that the radioactive substances have many uses in industry and medicine. Thus, for about the most hazardous substances imaginable, she suggests that instead of sequestering them *away* from access to the biosphere, we bring them right into our daily life. In essence, this amounts to a proposal that each of us can guard a bit of the radioactive waste ourselves, and use it while guarding it. Surely *that* will guarantee that we will not be harmed by the waste.

Dr. L. Douglas DeNike, a leading foe of nuclear power, has characterized the debates over "permanent" radioactive waste disposal as a tempest between the "Heap It and Keep It" band, and the "Chuck It and F--- It" group.

But actually, the nuclear promoters have founded at least two additional schools of radioactive waste

\+ + + + + + + + + + + + + +

(5) Ray, Dixy Lee, "Irrational Fears of Runaway Nuclear Energy Don't Stand Up against Scientific Evidence", *The Sunday Oregonian* (Portland, OR), Forum Section, October 26, 1975.

Editorial cartoon by Bill Sanders.
Reprinted courtesy of Field Newspaper Syndicate.

"Sorry about that, but 'a miss is as good as a mile' as we always say!"

management: The Pharmacist School and the Gourmet School.

The Pharmacist School

The pharmacist school has studied the problem very carefully, and has announced that each persons's share of the radioactive waste from a year's worth of electricity generated by nuclear fission, would be approximately the size of an *aspirin tablet*. Some have risen to challenge the *size* of the aspirin tablet, but I regard this as a secondary controversy.

Every pharmacist and fifth grader realizes that it is not the volume but rather the toxicity which matters. Each tablet could contain the equivalent of a million lethal doses of radioactivity. Moreover, participants in this tablet controversy conveniently overlook the fact that, in order to fuel just one nuclear power plant for one year, we must create a half-billion pounds of mine "tailings", which are *radioactive wastes.* The General Accounting Office of the Congress was *nearer* the mark when it testified that "by the end of this century . . . there will be one billion cubic feet of nuclear waste in the U.S.—enough to cover a four-lane highway coast to coast a foot deep."

Nevertheless, predecessors to the pharmacist school were glibly telling the public that a home garage could hold a good fraction of the radioactive waste. Still others argued that the entire nuclear industry's waste could be lodged in an area no larger than a football field.

Note well that irrelevancies (the physical volume of the waste) have been discussed in depth, but no consideration is given to the real problem, namely getting the *required fraction* of the waste into the aspirin tablet, the garage, or the football field.

The Gourmet School

The gourmet school of waste management is even more delightful. The two leading exponents of the gourmet school are Dr. Bernard Cohen and Dr. Dixy Lee Ray.

Dr. Cohen has published a long and sober paper in the *Scientific American* (6), with elaborate calculations to prove the following remarkable conclusion:

> "In fact, one can calculate that after 600 years, a person would have to ingest approximately half a pound of the buried waste to incur a 50% chance of suffering a lethal cancer."

Dr. Cohen did not state whether this was with or without worcestershire sauce.

Dr. Ray's gourmet tastes run to plutonium. She has written extensively about eating plutonium (7):

> "We are told repeatedly, however, that plutonium is the most toxic thing that has ever been brought upon the face of the Earth. Toxic compared to what, I ask. Remember, plutonium is a metal. It can be made into a powder by making it into an oxide. It is awfully hard to dissolve it, but if one takes plutonium oxide and dissolves it in water on an equal molecular basis with the same amount of molecules from a biological toxin, say botulism, the famous food poisoning, some interesting and surprising comparisons of toxicity can be made.
>
> "So what do you have in the way of a toxic tea? Well, one teaspoon of that botulism concentration is enough to kill 140,000 people and one teaspoon of the plutonium oxide would be about one half of one lethal dose for one human being."

I could point out to Dr. Ray that a *little understanding* of the plutonium problem would reveal to her

\+ + + + + + + + + + + + + +

(6) Cohen, Bernard L., "The Disposal of Radioactive Wastes from Fission Reactors", *Scientific American 236*, 6:21-31, June 1977. (The eating of wastes is discussed on page 28.)

(7) Ray, Dixy Lee, "The Case for Nuclear Power". Presented by the Western Environmental Trade Assn. of Washington, Seattle 98101. This pamphlet is undated, but appeared in late 1975 or during 1976. (The specific citation on eating plutonium occurs on page 4.)

that plutonium is about 100,000 times as effective in producing cancer when it is *inhaled* rather than *eaten*, but why should I be a spoiler for her gourmet interest in plutonium tea?

[Note: It now looks, moreover, as if she was sounding forth from somewhat faulty assumptions concerning the low amount of plutonium which can be absorbed through the gut; the September 15, 1978 issue of Science magazine carries new information on absorption of plutonium from chlorinated drinking water. Dixy Lee Ray may have to make her plutonium tea with the purest of spring water.]

These are illustrations of the lengths to which the nuclear promoters will go to obscure the real issue, namely the losses of radioactivity on the way to the bank.

However, I am *not* suggesting that containment within a final burial site is an unimportant matter. It is to people's credit that they are gripped by the depravity of reaching into the future with our radioactive garbage. However, it is false to think of *early* leaks of garbage and *later* leaks of garbage as morally different! Both kinds of leaks degrade future life on Earth. The biological consequences and the moral issues are the same whether the leaks originate from scattered locations on the surface of the planet or from a central burial ground.

What I am saying is that losses from the burial site are *secondary* to the losses which will inevitably occur *on the way* to the burial place. If we can't ever solve the problem of the losses on the way to the bank—and I am positively convinced that such a solution is not credible—then it is simply irresponsible to focus on the burial problem. We need to focus on *stopping the production* of additional radioactive poisons.

Losses: How Close to Zero?

Since we know how much radioactivity is produced per year's operation of a nuclear power plant, and

"My God, for a second I thought it said 'plutonium'."

since we know how many cancer fatalities will be produced per unit of radiation received by humans, the only remaining factor necessary to calculate the public health consequences of nuclear power is the fraction of radioactivity which escapes confinement.

With a little arithmetic (8), it turns out that for a full-fledged nuclear energy economy of 1,000 large nuclear plants, and for a containment perfection of 99.9% (meaning a loss to the environment of only one one-thousandth of the long-lived radioactivity), we will in time add roughly 198,000 extra fatal cancers *annually* in the USA alone for each year such plants operate.

It is doubtful that the public will stand still for 198,000 *extra* fatal cancers per year. I doubt that an informed public would stand still for 19,800 extra fatal cancers per year.

Let us assume that the public might accept 1,980 extra fatal cancers per year, which sets aside the large matter of additional *genetic* deaths at that level of irradiation.

To reduce the killing from 198,000 per year to 1,980 per year, the nuclear industry would have to increase its level of perfection from 99.9% to 99.999% in the control of its radioactive poisons. This means that no more than one part out of every 100,000 parts of radioactive garbage can be released to the environment!

It takes just a *little* sober consideration to appreciate how marvelous a degree of containment it would represent to lose only one unit of radioactivity out of every 100,000 units handled. And while making this consideration, we must take into account not only the planned releases during "normal" operations, but also all the burps, spills, leaks, and unexpected "migrations" of radioactive poisons

\+ + + + + + + + + + + + + +

(8) Gofman, J.W., "Statement in the Matter of 10 CFR 51, Licensing of Production and Utilization Facilities; Environmental Effects of the Uranium Fuel Cycle". Submitted to the Nuclear Regulatory Commission for Docket # RM 50-3, by the Sierra Club (Buffalo, New York), October 3, 1977.

which have continuously plagued the nuclear industry's operations throughout its history.

It is the sum total of losses in all of a very complex series of operations, including transportation steps, which would *have* to remain below one part in 100,000. Hardly consistent with containers which have leaked plutonium from a truck traveling through metropolitan New York City. Hardly consistent with the spreading of about 200 grams of plutonium into the area just west of Denver by the Rockly Flats Plant, while at the same time official records showed no losses of plutonium beyond a few *milli*grams. Incidentally, citizens and *not the officials* discovered the leak of plutonium.

Perhaps you have noticed that every time a radioactive release is known to have occurred, officials announce, "But the amount released poses no danger to public health". There must, by now, be 100,000 such announcements. How many "small" releases can we have and still have the total "small"?

The Crucial Assumption Should Make Sense

It defies all common sense to believe that containments "on the way to the bank" are going to be as good as 99.999% perfect. I think, as a chemist, and with a long association with the nuclear enterprise, that if the industry achieves a level of 99.9% perfection, it will still be a miracle. And yet that miracle would be far from good enough—it would be, unequivocally, a public health disaster.

[*NOTE:* By no stretch of the imagination could anyone assert that radioactive garbage buried in plastic sacks or rustable barrels in shallow trenches is "contained" or isolated from the environment. A great deal of dilute waste, however, *is* buried that way. But where? The Department of Energy

Calvin Grondahl in the *Deseret News*, Salt Lake City.

(successor to the Atomic Energy Commission) is now pleading with the public to help FIND its buried radioactive waste: ''DOE has turned to the public for help in identifying all possible sites in the U.S. where radioactive material may have been stored or processed . . . . Because many records were misplaced or destroyed over the years, the Department has asked that any member of the public who knows where such work was once done contact DOE. The sites would have been used for nuclear work from the 1940's through the 1960's, but might be used for non-nuclear purposes today.'' From DOE Office of Consumer Affairs, May 25, 1978.]

Operating Without the Foggiest Notion

You might well ask, ''Don't the officials know what their containment ability is?'' I've already described for you how blissfully ignorant the officials at Rocky Flats were, when they were wrong by a factor of some 20,000-fold in their estimated plutonium losses.

But I can generalize. The straightforward answer is that neither the nuclear industry nor the governmental regulatory agencies have the foggiest notion of how well or poorly they are doing at their containment task.

One point is absolutely certain: their *measurement ability* falls far, far short of what it would take to know that they could prevent more than one part in every 100,000 parts from escaping into the environment on the way to the bank. Of course, not having the evidence has never impeded such agencies as the Nuclear Regulatory Commission from making extravagant *claims* about how well they are going to contain radioactivity in the future.

From the comfort of nice offices, they figure out what containment would be desirable, put numbers into a computer, adjust the program until it produces numbers which they like as output, and then they publish these figures in mammoth reports about hypothetical containment.

If asked why these numbers seem out of accord with performance to date, they promptly announce that past experience is part of a "learning curve", and that they will do far better in the future. That their numbers have absolutely nothing to do with past experience in the real world, or with the likely *future* experience in the real world, is of little moment. After all, Grimm "made it" in a big way writing fairy-tales, and who is to say that the Nuclear Regulatory Commission will not outdo Grimm in the latter half of the twentieth century?

Near Perfection Required

The containment of the so-called fission products, such as the familiar strontium-90 and cesium-137, to only one part lost per 100,000 would indeed be a prodigious achievement. I doubt that any rational human, using simple common sense, would give credence to engineers and scientists who blithely assure that they will be able to do it. But difficult as this feat would be, it is *not* the MOST difficult feat required of nuclear engineering. What is the more difficult feat?

It is the handling and containment of the "alpha-emitting" radioactive substances, which are a separate class from the fission products. I will explain.

The current light water reactor program is clearly a dead-end. There is not enough suitable uranium ore available in this country for any expansion providing significant amounts of energy. So, no matter what is done about the Clinch River Breeder Project, any real use of nuclear power will indeed require *some* type of breeder reactor, if one can be technically made to work. Either such a reactor will be based upon the uranium-plutonium cycle, or upon the thorium-uranium-233 cycle. Any assurances uttered that we are *not* going into such cycles, would be the same as saying we should abandon nuclear power right now.

If the current light water reactors are the only type which are ever going to function, their trivial energy contribution is indeed ridiculous. With the equivalent of 50 large nukes operable today, their combined contribution to our energy supply is about 3%; of course the percentage is about 4-fold higher if just *electrical* energy is discussed. Only 140 nukes can be fueled with our domestic supply of uranium, and they could not supply more than 10% of the country's gross energy (9).

So the nuclear future depends upon breeders. And the President has made it very clear that he is enthusiastic about breeders. He has called for a massive outlay of tax-dollars for breeder research. He has said explicitly that plutonium breeders may emerge as an integral part of his energy program. His only reservation is that he would want to introduce the breeder commercially at that mythical time when nations no longer want nuclear bombs!

Whether the breeder program is based upon plutonium or whether it is based upon uranium-233, we are going to produce and handle these substances by the thousands of tons. These two substances are in the class of alpha-emitters, providing the same sort of irradiation which is already responsible for the epidemic of lung-cancer deaths claiming the lives of our uranium miners on the Colorado Plateau.

As with our earlier arithmetic for the fission products, we can calculate the number of tons of these substances which will have to be handled, and we have good estimates of the number of lung cancers to be expected per unit quantity of such substances inhaled by people (10). From those figures, we can calculate the degree of containment which will be *necessary* to prevent a massive epidemic of lung cancer in this country.

+ + + + + + + + + + + + + + +

(9) Gofman, J.W., "Gross Energy Available through Light Water Reactors", *Committee for Nuclear Responsibility Report 1977-2*, May 1977.

(10) Gofman, J.W., "The Plutonium Controversy", *Journal of the American Medical Assn.* (*236*), 3:284-6, July 19, 1976.

Gofman, J.W., "The Cancer Hazard from Inhaled Plutonium", *Congressional Record* 121:S14610-6, 1975. This is also a C.N.R. Report, May 1975.

It turns out that the required degree of containment for such substances as plutonium or uranium-233 is ten times *more* severe than for the fission products we have already reviewed. Containment of all but *one part per million* is required to avert a public health disaster. Thus, all the steps combined must achieve containment to 99.9999% perfection.

Muffing It

Yet routinely, the industry loses track of its alpha-emitters. The term for lost plutonium and uranium is "MUF"! "Material Unaccounted For", or sometimes "Inventory Differences". DOE provides a soothing description of these ominous terms: "These terms relate to the differences between the amount of material which bookkeeping entries indicate is present, and what a physical inventory shows to be on hand at the end of an accounting period. Since the amount of material shown on the books and the amount on hand are subject to measurement error, the two figures are seldom identical, thus creating an inventory difference."

In the past, MUF has averaged 1% of the inventory. Due to conveniently changing terms, it is hard to say what it is currently. Suppose MUF were only 0.1% of inventory. That would mean they can't find 1 part per 1,000 . . . yet they are assuring the public that they *can* and *do* contain all but 1 part per 1,000,000. Simply a fraud.

I am sure you are aware that these particular alpha-emitters—plutonium and uranium—are the stuff out of which atomic bombs are fabricated. So, over and above the potential health consequences, the national security consequences of losing such materials are not taken lightly. Yet recent investigations show that for materials in this class—so-called "special nuclear materials"—some four tons or so cannot be accounted for by the processors.

Not 4 grams. Not 4 pounds. But 4 *tons*. Enough to make 400 atomic bombs. When asked about the missing tons, the governmental agencies answer that the missing

Oliphant, © 1977 Washington Star-Los Angeles Times Syndicate.

material must be stuck in the pipes. Or somewhere. *Of course* it has not escaped . . . .

As for their knowledge that the material can't have escaped, I should repeat that their measurement abilities fall far, far short of being able to know *anything* about the losses. And these are the same people who are assuring us they will, in the future, accomplish the miracle of containing 99.9999% of such materials.

We have some interesting benchmarks with which to assess the remarkable capabilities of the nuclear industry. So good is its ability to contain plutonium, that two grams of it recently fell out of a *filing cabinet* at the Rocky Flats plant. If two grams can wander into a filing cabinet without being missed, do you believe it is likely that only two grams will be released per year by the entire nuclear fuel cycle supporting 240 nukes? Yet *that* is the paper prediction of the Environmental Protection Agency, blood-brother to the Nuclear Regulatory Commission. EPA loves computers too. With a minor adjustment in program, it can have a new set of numbers to meet any political requirement within 30 seconds. Why bother with reality?

Thin Air and Quicksand

There is a professor of nuclear engineering whom I have debated several times. Not too long ago, I showed him some of my work which led to the conclusion that the toxicity of plutonium had been seriously underestimated. My conclusion is that the plutonium already dispersed from atmospheric bomb-testing has signed the death warrants of some 1,000,000 persons in the northern hemisphere, and the estimate covers only the deaths which will occur over the first 30-50 year period (11).

+ + + + + + + + + + + + + + + + + + +

(11) Gofman, J.W., "Estimated Production of Human Lung-Cancers by Plutonium from Worldwide Fallout", *Congressional Record* 121:S14616-9, 1975. This is also a C.N.R. Report, July 1975.

I told this engineer that the containment requirement for plutonium handled in the power program would be much more severe than had been previously assumed. "John," replied my professor friend, "just tell us what is needed! One part per million, one part per billion. You name it, and we'll engineer for it!"

What reasonable humans, using common sense, would give any credence *whatever* to other humans who claim they will contain 999,999 parts out of every million of plutonium they have to handle? And 99,999 parts out of every 100,000 parts of the fission products?

I find these claims to be utterly absurd on their face, irresponsible in the extreme, and based on nothing but thin air. Yet they represent the quicksand on which the nuclear industry is based. They represent the crucially different assumption which nuclear advocates *have to* make, either explicitly or tacitly, in order to justify nuclear power. Sad to say, they represent the kind of fantasy which shows that some sincere nuclear advocates have lost touch with reality.

Horse is Leaving the Barn

As I stated earlier, the reason the public has not reacted to the absurdity of claims that the required degree of containment will be achieved, is that its attention has been cleverly diverted to the question of where the "permanent" waste repository will be located. But by the time the burial ground is reached—wherever it is—this horse will long have been out of the barn, irretrievably, and the epidemic of diseases will be with us, and for millenia.

If the public really knew that the larger problem by far is what happens on the way to the bank, we might have some real public reaction to the nuclear power program. But I can assure you, with no fear of contradiction, that neither the government nor the industry is going to help the public understand the problem.

Courtesy of Jules Feiffer.

I want to leave you with what I consider the overwhelmingly important issue regarding nuclear power. It is that our health and lives are indeed threatened by it, and that our decision will affect virtually all generations to come. The health threat arises because the required degree of *pre-burial* containment of fission-product garbage, of plutonium, of uranium-233, *is never going to be achieved*.

The *pre*-burial containment problem is the Achilles Heel of the nuclear industry, not the burial problem itself. And the pre-burial problem is sufficiently severe to make it correct to say that the radioactive waste problem will *never* be solved. It won't be solved because the containment requirements "on the way to the bank" are far beyond any rational human expectations.

If we do not become concerned about those losses on the way to the bank, they will do us in. Far better that we understand the problem, and do nuclear power in first.

□ □ □ □ □

4.

THE CATCH-22 SOCIETY:

Some Thoughts on "Civil Disobedience" and "Uncivil Obedience"

There are two fundamental statements I would like to recall to your attention. The first is from the founding fathers in that long-forgotten document called the Declaration of Independence:

> "We hold these truths to be self-evident, that all men are created equal, that they are endowed by their Creator with certain inalienable Rights, that among these are Life, Liberty, and the pursuit of Happiness . . ."

Not at all bad! The only difficulty is that the history of our country shows that these words have been honored almost wholly in the breach. Today, if you seriously suggest you believe in the Declaration of Independence, you are far more likely than not, to get a dossier in the files of the FBI, CIA and the Justice Department. Now how did this come about, from such an auspicious start?

Let us turn to the second fundamental statement, a bit more down to Earth, from someone who told it "like it really was"—the former slave Frederick Douglass.

On August 4, 1857, Frederick Douglass said:

> Those who profess to favor freedom, and yet depreciate agitation, are men who want crops without plowing up the ground. They want rain without thunder and lightning. They want the ocean without the awful roar of its waters. This struggle may be a moral one; it may be a physical one; or it may be both moral and physical. But it must be a struggle.
>
> "Power concedes nothing without a demand. It never did and it never will. Find out just what people will submit to, and you have found out the exact amount of injustice and wrong which will be imposed upon them; and these will continue until they are resisted with either words or blows, or with both. The limits of tyrants are prescribed by the endurance of those whom they oppress."

Some may think that Frederick Douglass was referring to some distant society, from other times, where tyrants wore a crown. But tyranny comes in several forms, which Douglass appreciated very well indeed, and it comes in the form *we* know, with all the trappings of democracy.

Screwers and Screwees

In terms which are more in keeping with the parlance of today, I would paraphrase parts of Douglass' statement as follows:

> There are two fundamental groups in society, the screwers and the screwees. The screwers have all the apparatus of power, the sycophantic henchmen who do their bidding, and they have unflinching devotion to the preservation of their privileges at the expense of the screwees. Further, the screwers have enormous difficulty understanding why the screwees should ever raise any questions about the super-wonderful system which they have in place.

Generally speaking, screwees have never particularly enjoyed being overtly known as such. Therefore, a subterfuge is essential. The subterfuge which has emerged is the myth that the screwees are the ones who are really running the show, and that they do so through a democratic government. The ostensible purpose of government is to protect the rights and security of its citizens. This is done through a system of laws, drafted by hordes of those individuals we call lawyers, such laws being so written as to defy comprehension by virtually anyone, but never written so as to be *neutral* in any conflict between the screwers and the screwees.

Cartoon by Ken Mahood.

Remember, Rupert, until the meek DO inherit the earth, they are there to be trodden upon.

There is not necessarily any desire to be evil on the part of the screwers. All they want is an absolute guarantee that they can preserve and extend their privilege at the expense of the screwees. Stated otherwise, they wish to acquire an ever-increasing share of all the means of production and resources of the Earth, so that they can *still further* increase that share. And to these ends, we have the so-called "economy", which through ceaseless churning, steadily allows those with power and privilege to increase both. Thus the top 19% of families owns about 76% of all the privately held wealth in the USA, while the bottom 25% has no assets at all (Dr. L.C. Thurow, M.I.T. Department of Economics). The concentration of wealth and power is such that recent estimates are that the top 5% of the American population owns more assets than does the bottom 81% combined (also Thurow).

What is manufactured in this "economy" is really quite irrelevant to the screwer-class. The only criterion is that what is manufactured be saleable at a profit. Hula hoops, arms, oil, cars, cigarettes, nuclear power plants, food, are all viewed through only one lens—can they be sold at a profit. Better still are those products which, through built-in obsolescence, can insure that the purchaser becomes locked into the system of dependence. Best of all are those products which become absolute necessities in the contemporary way of life, and which cannot possibly be made by the screwee himself. Thus, for example, nuclear power plants to create electricity are lovely, whereas small solar systems are a disaster—from the point of view of the screwers.

Technology vs. Inalienable Rights

I stated above that the screwers are essentially neutral on such issues as the health of people, preservation of a livable environment, beauty of the landscape. They couldn't give less of a damn about such issues. Their actions here and abroad prove they regard people and the environment as expendable. As for people, more can be created, and besides, the planet presently has a surplus of these potentially trouble-making creatures. As for the environment, one simply moves on to a *new* location.

This neutrality exists so long as the cost to their enterprises is inconsequential. But let someone raise the issue that pollution of the environment, poisoning of people, poisoning of other life forms, should be prevented and that the responsibility for such prevention be borne by the screwers, and all hell breaks loose. Or if someone suggests that a particular profitable activity should cease entirely because it *cannot* be made safe, we are in for a battle royal. They tell you that Kepone and dioxin put bread on your table.

Ostensibly, those who are concerned have a marvelous recourse—namely the government of, for, and by the people, which will (of course) be ever watchful of the interests of the people. The Congress will pass the laws necessary, the Executive Branch will set up administrative agencies to flesh out the programs and regulatory agencies to insure that the public is protected from depredations. If the laws don't seem to be achieving the goals dreamed of by the screwees, they can turn Tweedle-dum out of office and elect Tweedle-dee. When Tweedle-dee turns out to be a carbon copy of Tweedle-dum, then there is always the privilege of electing Tweedle-dum's cloned brother.

So long as the momentous issue is whether we shall have green Frisbees instead of white Frisbees, no one is particularly exercised. But, unfortunately, with the advent of modern technology, the issues become much more serious. Modern technology can very definitely impact seriously and clearly upon health, upon the quality of life, and upon the very inalienable rights discussed in the Declaration of Independence: Life, liberty and the pursuit of happiness.

Technology "Pushers" Who Will Fight to Your Death

Modern technology has some other properties of importance, namely that it is generally large-scale and very expensive. The screwers rarely like to take risks these days, with the demise of the free enterprise system, and they prefer to have the screwees pay the costs of exploration of new technologies through taxes. This is called stimulation of the economy and of the private-sector by government. So we establish huge bureaucracies in the government to stimulate the new enterprise, through gifts (hidden and not-so-hidden) from the taxpayers to the corporations engaged in the technology.

On top of this, we create huge regulatory bureaucracies to see that all's done fair and square. Now—instead of just the corporate interest in the venture—we have two new and huge interests: (1.) The enterprise-stimulators, as for example, the Department of Energy (formerly ERDA, and before *that*, the Atomic Energy Commission), and (2.) Innumerable protection agencies, or regulatory agencies, such as the Nuclear Regulatory Agency, the Food and Drug Administration, and the Environmental Protection Agency. And all three of these very highly vested interests will fight to your death to preserve and extend a new technology, and to extract profits from it.

Nuclear energy is only one example, but since we are discussing the monstrosity known as the Barnwell Nuclear Fuel Reprocessing Facility today, it is appropriate to explore some deeply important issues by utilizing nuclear energy as our center-piece. Our concern here today is not technology *per se* so much as it is how we can cope with this *and other* monstrosities foisted upon society, in the name of privilege preservation and extension. By the way, I am not suggesting that *all* immorality resides within the screwer-class! It's simply that the privilege-elite has such extraordinary POWER—which magnifies its immorality—whereas the screwees do not.

The Three Nuclear Products

Starting with the development of atomic weapons in World War II, we created the circumstances for establishment of one of the "stimulation" bureaucracies, called the Atomic Energy Commission. Once established, the chief function of the AEC was its own self-perpetuation, needed or not. The AEC had three products, in essence: (1.) Bombs made of heavy elements undergoing fission very *fast*; (2.) Nuclear reactors made of heavy elements undergoing fission *more slowly*; and (3.) a huge and growing inventory of radioactive poisons. The way to preserve and expand the bureaucracy was to increase production of all three and (as rapidly as possible) to provide subsidies, tax breaks, and other goodies to entice corporations to join the venture in several ways.

As for the bombs, the "business" is doing very well because we have created a dynamic with the Soviet Union which can hardly be broken.

Another aspect of the bomb business—blowing up tens of thousands of nuclear bombs everywhere in the so-called peaceful bomb program (or "Plowshare Program")—has not fared quite so well recently. The AEC was gung-ho just a little while ago to divert rivers, cut passes through mountains, and to build a new Panama Canal, with hundreds of megatons of hydrogen bombs. It was going to help detonate some 15,000 nuclear bombs under the Rocky Mountains! *That* was to be for the stimulation of natural gas production, but the gas from the first few experiments was contaminated with radioactivity, of course. It was not the biological insanity which stopped the Plowshare projects momentarily, but rather the need for some detail work to insure that too many of the "commercial" bombs did not get into the wrong hands during the development of this super-genius idea.

As for the product called waste—meaning monumental quantities of radioactive poisons—sales are not doing too badly, with more and more of the garbage finding its way into our industrial economy, into medicine, and into our homes and hearts, via the ionization-type smoke detectors, pacemakers, and even artificial hearts. If we follow the advice of Dixy Lee Ray and Ronald Reagan, we will eliminate the problem of radioactive waste by bringing it into our homes and other activities, and re-naming it to be a "resource". They both foresee tremendous growth for this part of the business. Ronald Reagan is telling his readers (Ronald Reagan Speaks Out, April 1, 1978) that the market for plutonium-powered hearts may be 182,000 *per year*. If the Nervous Nellies can be kept under control, presumably. *

The third product—nuclear reactors to boil water—is one of the real darlings of the AEC and of its successors by other names, and has attracted the dollars of certain corporate giants, including the electrical utility

Courtesy of Geoffrey Murphy and Minnesota PIRG *Statewatch*.

\+ + + + + + + + + + + + + +

* The idiocy of plutonium-powered hearts is explored in the novel, *Heartbeat*, by Eugene Dong, M.D., and Spyros Andreopoulos. New York: Coward, McCann, and Geoghagen, Inc., 1978.

industry and the banking business. Nuclear reactors are NOT a good way to help cope with our energy needs, but that fact is highly secondary and even irrelevant, because they keep the bureaucracies in business (including the *regulatory* bureaucracies), and they are expected to make profits for the corporations which mine and mill uranium, and which construct and operate the plants associated with the whole enterprise.

Gina Lollobrigida's Advice

But some clouds have darkened the horizon of the atomic energy wonderland. Even back in the 1950's, the St. Louis Citizens' Committee and also Professor Linus Pauling began to point out correctly that it likely would be atomic energy or us, *but not both*.

The governmental agencies, like the AEC, the Federal Radiation Council, the National Academy of Sciences, the EPA, the Nuclear Regulatory Agency, and the ERDA, all have pitched in with gusto in the effort to insure three things:

(1.) That the hazards be declared non-existent;

(2.) That any research which might uncover *the real hazards* be gutted, and that the investigators be banished into limbo if they so much as suggest that radioactivity kills people through cancer, leukemia, and genetic diseases;

(3.) That the public (that is, the screwees) be constantly reassured that there is no way the monstrous amount of radioactive garbage is ever going to reach people and poison them.

The only trouble with these agencies is that they have lied so frequently and covered up so often, that they can't even begin to keep track of their lies. There may be, somewhere in the hinterlands, some gullible souls who still believe what these agencies tell them, but certainly there is no rational reason to believe anything they say. I've often wondered why these agencies did not heed Gina Lollobrigida's credo, when she said she never lies simply because it's just too damned hard to keep track of the lies.

To the best of my scientific ability, I have considered the question of whether there is any way that man can live a healthy life with nuclear energy, and I have reached the conclusion that he cannot. The requirements for adequate containment of the radioactive by-products are far beyond any REASONABLE expectation for human endeavor, and the result of failure is an inevitable epidemic of additional cancer deaths, leukemia deaths, and genetic deaths, which are going to plague humans for *generations*,

Marian Wolff for Truth in Power.

even if the monstrous stupidity known as nuclear energy were to stop within the next few decades.

I have examined the arguments of the promoters of nuclear energy, and they always boil down to the same absurdity: "If everything goes *perfectly*, then everything will go *perfectly*."

Or, "Trust us! Even though we have come close, we still have our first major city to knock over."

But I am far, far, from alone in these concerns. Today, I would make the guess that most thinking Americans share my concerns—that is, the people outside the fairyland of atomic energy paychecks—and they want to do something about these grim prospects for themselves and their children. Even those who answer pollsters in the affirmative about the development of nuclear energy *far more often than not* suggest that the plants be built elsewhere, anywhere, but not near THEM. There's a ringing endorsement for nuclear energy.

Self-Defense, Random Murder, "Good Germans"

So we come to the really important questions we should be considering today. Among these questions are the following:

(1.) Even assuming we are perfectly willing to be shafted economically for the benefit of privilege-extension and profits, do we not have any RIGHTS OF SELF-DEFENSE against becoming a victim of the forthcoming cancer plague being so generously given us by government and industry, in cooperation? Do we not have any rights (really obligations) to defend our children and the generations yet unborn from these plagues upon their health?

(2.) If the *majority* of the public, through gullibility, fear, and outrageous propaganda campaigns, is convinced that a technology like nuclear energy should *go ahead*, does that justify permitting that technology to COMMIT RANDOM MURDER on thousands or millions of members both of the majority and the minority? The Aztecs once made large numbers of human sacrifices to appease the Gods. We shudder about that, but how does it differ from the human sacrifice, essentially by drawing lots, which we will create on a far grander scale to appease the privilege-seekers, who threaten us with economic extinction unless we are compliant? Would we be willing to accept nuclear power if we had to name 100 or 1,000 or 100,000 people each year to be executed by firing squad, in exchange for electricity? Is this really different from what we are accepting, if we permit nuclear power to go ahead?

(3.) What shall we think about ourselves as we become party to premeditated random murder, to the destruction of life, liberty and the pursuit of happiness? Just recently, as a result of the NBC programs called "Holocaust", Americans have had the opportunity to point the finger at all the "good Germans" who did nothing to prevent the murder of 11,000,000 humans under the Nazis. Just how do the 98% or more of us differ from the "good Germans", as we participate every day in the same sort of holocaust, even if the new holocaust is more subtle than gas chambers or the firing squads of Babi Yar? And assuredly WE DO PARTICIPATE IN SUCH HOLOCAUSTS by permitting nuclear power to grow, by permitting the technological race for a first strike in nuclear warfare, and by supporting the "leaders" around the world who murder and starve people by the *hundreds of millions*. We support those rulers in the name of security alliances—but often truly to provide slave labor for our multinational corporations, while at the same time the corporations throw an ever-increasing segment of the American labor force on the junk heap of human castaways.

Every one of us (and that *is* the 98%) who pays taxes to our government to support nuclear power, to support unspeakable regimes around the world, to sponsor technologies and other activities (like the Vietnam War) which violate our deepest ethical and religious beliefs—every one of us is really just the same as a "good German". Part of our labor each year is devoted to the payment of taxes for the committing of atrocities *little different* from those of the Germans, for activities we abhor.

". . . In affirming that a man may not be taxed unless he has directly or indirectly given his consent, [we affirm] that he may refuse to be so taxed . . . Perhaps it will be said that this consent is not a specific, but a general one, and that the citizen is understood to have assented to every thing his representative may do, when he voted for him.

"But suppose he did not vote for him; and on the contrary did all in his power to get elected some one holding opposite views — what then? The reply will probably be that, by taking part in such an election, he tacitly agreed to abide by the decision of the majority.

"And how if he did not vote at all? Why then he cannot justly complain of any tax, seeing that he made no protest against its imposition.

"So, curiously enough, it seems that he gave his consent in whatever way he acted — whether he said yes, whether he said no, or whether he remained neuter! A rather awkward doctrine this."

Herbert Spencer's
"The Right to Ignore the State", 1844,
in his book, *Social Statics*.

The Grievance Process!

We are treated to a remarkable spectacle. If we don't like what is being done in our name and with our dollars, we can change things through law, by electing Tweedle-dee instead of Tweedle-dum. If we object to the activities of the Atomic Energy Commission or the Nuclear Regulatory Commission, we have the fabulous privilege of "intervening" in license-hearings. Citizens are expected somehow to hire lawyers in such processes, while their tax dollars go to support an *army* of lawyers at the beck and call of the Nuclear Regulatory Commission. There has not existed the slightest shred of meaningful evidence that the entire intervention process in nuclear energy is anything more than the most callous of charades and frauds. Short of direct proof that a nuclear reactor is sitting on Mount Vesuvius at the height of its eruption, there is little doubt that the Nuclear Regulatory Commission will approve the site. Probably some of the Commissioners would suggest coming back next week . . . maybe the volcano will quiet down.

All this failing, we can go to the courts for relief! And there—just as in the interventions—we can pay lawyers out of non-existent citizen-funds, while our tax-dollars and subsidies to corporations enable *them* to hire the best legal talent in the country to prevent a fair hearing before the courts. Some environmentalists hail the precious *few* decisions which have come down on their side, but in the main, just about nothing has been accomplished in the courts to stop the nuclear power program.

Yes, there is the magnificent decision of Judge MacMillan (North Carolina) declaring the unconstitutionality of that piece of Congressional chicanery known as the Price-Anderson Act, which *limits the liability* of the nuclear industry in case of accidents.

But I will predict that either the Supreme Court will find a way to accommodate the nuclear industry, or the Congress will develop a *new* bail-out protection-program for the benefit of the nuclear industry. That industry simply is NOT willing to operate nuclear power plants if its members have to be liable for the murders they are assuredly committing and are going to commit, for the farm produce they are going to render unsaleable, and for the land which is going to become uninhabitable—possibly for decades—after a major nuclear "incident".

[*Note:* In late June, 1978, the Supreme Court did reverse Judge MacMillan, and declared the Price-Anderson Act to be Constitutional.]

Protection by 50,000 Coconut-Heads

Farmers should take particular note of the problem which nuclear power presents them. The Atomic Energy Commission, with all its talent and equipment, pronounced Bikini Island fit again for the natives to return. The coconuts at AEC headquarters planted 50,000 new coconut trees on Bikini because radioactivity in the soil was *no problem* there. A little later, the AEC said the natives had better not go too strong on eating coconuts. Then it was down to recommending ONE coconut per day. And finally, none! It has just been announced that the islanders must again be moved off, because Bikini is "unfit for human habitation and may remain so for decades". There's real competence . . . and re-assurance . . . and 50,000 coconut trees. Farmers should really feel good about the assurances concerning the Barnwell plant and nuclear reactors, with those assurances coming from the very same coconut heads.

Screwees' Choice:

Finally, the citizen has the privilege of non-compliance in one or more ways. Some have tried tax-resistance rather than complicity with moral outrages committed in their names. Either the funds are seized by Internal Revenue—which, of course, makes the protest only a gesture—or the individuals may go to jail. There are a few *successful* tax-rebels, however, and we may wish to learn from them.

Then there is the privilege of "civil disobedience". Our great moralist President has spoken out on this privilege. He says that the citizen has the right to civil disobedience provided he is willing to pay the penalty, that

From *Cobb Again*, Wild and Woolley, Sydney, Australia.

From Ireland.
By Martyn Turner in *Nuclear Ireland* (by M. Hussey and C. Craig, Irish Writers' Co-op, 50 Merrion Sq., Dublin 2). Originally from the *Irish Times*.

* Ireland's Minister for Industry, Commerce, and Energy, Desmond O'Malley.

is, go to jail. It seems somewhat curious that a President who says nuclear power is so dangerous that it should be used as a last resort (during his campaign) becomes a President with a gung-ho, full-speed-ahead nuclear program, and gets $250,000 pay per year for this fulfillment of a campaign promise. He thinks it is perfectly all right for you to object to nuclear power with "civil disobedience" if you are willing to go to jail. A very peculiar set of standards about who deserves pay and who deserves jail.

Civil Disobedience vs. Uncivil Obedience

We most certainly have a curious set of standards in general about what constitutes "civil disobedience." I would have to presume that the desired behavior is "uncivil obedience". "Uncivil obedience" is that wonderful state where you declare yourself to be a moral cripple, a *de facto* slave contributing your labor and dollars to outrageous offenses against humanity, or both. Let us consider some of the prime examples of "uncivil obedience".

For a period of over 20 years, the Atomic Energy Commission and its successors, the Energy Research and Development Administration and the Department of Energy, have knowingly and willfully deceived the American public about the hazards of low-dose radiation. Their officials have lied consistently, they have suppressed reports about hazards, they have suppressed scientists who tried to tell the truth about radiation hazards, and they have used public funds to mount an unconscionable propaganda campaign about "clean" energy to our impressionable children in the schools.

They have not been arrested for "civil disobedience", although it is truly hard to understand why they should *not* be in jail. Instead, for their courageous acts of "uncivil obedience", they have all moved up to even more comfortable sinecures in the Department of Energy, with our tax dollars to pay them for these twenty years of defrauding the public. I guess there really isn't any *law* against willful deception of the public. I do seem to remember a few years ago an impassioned statement in the newspapers by a government official who said he saw *no reason* why the government should not lie.

Most of you have recently read about the suppression of the work of Dr. Thomas Mancuso, who is confirming precisely what we knew nine years ago would *inevitably* happen—namely that workers receiving the so-called "safe" radiation dose at Hanford are indeed experiencing an increased risk of dying of cancer.

Dr. Mancuso was soon given *the axe* by Dr. Jim Liverman of DOE, who thereby is prolonging the deception of the public about low-dose radiation. After Liverman committed his courageous act of "uncivil obedience", the other two Jimmies—Carter and Schlesinger—announced his promotion to Acting Assistant Secretary of the Department of Energy (DOE). The only treat we have been spared is the familiar press conference with the Jimmies, Carter and Schlesinger, embracing the third Jimmie, Liverman, and saying, "Jim, we have full confidence in everything you are doing and in your courageous acts of 'uncivil obedience'." It is not that Dr. Liverman's actions are unknown to the President and the Secretary of Energy, for through the persistent and hard work of Robert Alvarez of the Environmental Policy Center, this scandal has been fully exposed in Representative Rogers' Congressional Hearings.

There was also ERDA's Dr. Robert Thorne. According to the General Accounting Office of Congress, Dr. Thorne misused public funds to spread misleading propaganda into California, in the effort to defeat Proposition 15 (the Nuclear Safeguards Initiative of 1976). For *his* courageous act of "uncivil obedience", did Dr. Thorne have to go to jail? Indeed not. Instead, he has been nominated by the two Jimmies to be an Assistant Secretary of Energy.

"Find Out Just What People Will Submit to . . ."

There is something especially galling about the idea of rewarding a respect for life, for human rights, for refusal to participate in moral outrages, *with a jail sentence*—and at the same time rewarding duplicity, connivance, fraud, and deception *with a high government salary*. But that is, of course, precisely the expectation if we really understand what Frederick Douglass told us over 100 years ago:

> "Find out just what people will submit to, and you have found out the exact amount of injustice and wrong which will be imposed upon them."

By permission of Community Press Features.

That nuclear power promotion is a human depravity of the first order has been evident to a limited group for some ten years. This is now becoming rapidly appreciated by an ever-growing segment of the screwee public. Suppose we win a ban on nuclear power because it is becoming evident that it is *also* an ECONOMIC crock. How much will we have won? As bad as nuclear power is in human terms for this and future generations, let there be no doubt that a privilege-extension and profit-oriented society will think up other activities *as bad or worse*. Anything goes, if a profit can be made from it, or if the dependence of the screwees can be increased.

Bringing Home the Nuremberg Principles

Tomorrow there will be worse things than nuclear power and the Nazi Holocaust, unless we can correct the bankruptcy of the average person (ourselves of course) with respect to being a *de facto* slave, a moral cripple, both being the essence of the so-called "good German" complex.

I think a major task for those assembled here is to give the deepest and most serious thought to taking away the power which others now have to force us to commit the most outrageous acts upon our fellow men at home and abroad, to force us to go to jail to defend the inalienable rights in the Declaration of Independence, to refuse us the opportunity in the courts to use the issue of a higher law in defense of so-called "civil disobedience", to force us all to be *de facto* slaves, moral cripples, "good Germans" in an on-going holocaust, overt and covert, both against this and future generations.

Let's face it. Either we learn to cope with that power *effectively*, or we remain the "good Germans" the vast majority of us are.

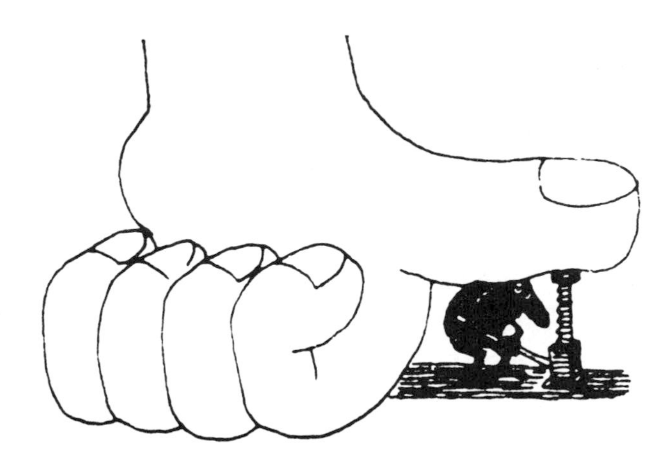

The anti-nuclear movement could become the driving force in spreading the Nuremberg Principle throughout the land—namely, that people will be held personally responsible for what they do to other humans. We tried the Nazis for crimes against humanity, even though many claimed that they were just following orders. We held them personally responsible for their acts. In the Nuremberg Principles, we stated explicitly the following:

- "Murder, extermination . . . or other inhuman acts done against any civilian population" constitute a crime against humanity;
- "Crimes against international law are committed by men, not by abstract entities";
- that a superior order "does not relieve a person from responsibility under international law, provided a moral choice was in fact possible to him"; and lastly,
- "Individuals have international duties which transcend the national obligations of obedience imposed by the individual state".

If the Nuremberg Principles were applied in our courtrooms, it would mean that those who refuse to stand by "like good Germans" while random murder by radiation is planned—indeed, who refuse to be *accomplices* in this program via their taxes—these defendants would be tried before juries who would have the right to rule that such defendants were *not* breaking the law, because they were obeying a higher law.

In my opinion, the anti-nuclear movement would have an impact far, far beyond the nuclear power issue, if it would insist that the Nuremberg Principles be applied throughout the land. Until that happens, what restraint is there on greater and greater government-licensed abominations? Until the Nuremberg Principles are established in this land, laws can force us to violate our deepest religious or ethical beliefs—making a travesty of freedom.

The way I see it, the significance of this rally and of future efforts to stop the Barnwell Plant is this: You are really in the vanguard of an effort to establish a major new principle in human society, a giant advance for justice and freedom. Keep thinking BIG, keep your eye on the ball, and don't give up! No nukes, y'all.

□ □ □ □ □

5.

A SMALL AFFIDAVIT WITH BIG IMPLICATIONS

"Crime does not pay at your level."

Drawing by Handelsman;

□ □ □ □ □

Note:

While I am not an *advocate* of ''occupations'' of nuclear facilities as an effective way to stop nuclear power, I am certain that simple justice requires that those who have been arrested for occupations, be allowed to explain to juries *why they did it* and how a ''higher law'' was involved in their action.

This brief affidavit was written at the request of some members of the non-violent Trojan Decommissioning Alliance who were arrested in November, 1977, for trying to shut down the Trojan nuclear power plant by blocking access to it. They were charged with criminal trespass.

The defendants, with the aid of the Community Law Project in Portland, tried to invoke Oregon's ''choice of evils'' law and to present expert witnesses to help them explain the consequences of nuclear power to the jury. However, the judge did not permit them to do so, and on June 30, 1978, the jury declared them guilty.

The affidavit refers to Exhibit 1, which is provided following the affidavit; however, Exhibits 2, 3, 4, 5, 6, and 7—which are technical papers—are not provided.

JWG.

□ □ □ □ □

Affidavit

DR. JOHN W. GOFMAN being duly sworn, deposes and says:

. . . As a result of my education and research in relevant areas, I feel qualified to make the statements which follow in this affidavit. A detailed biography is attached as Exhibit 1.

If called upon to testify in the trial of the Trojan Decommissioning Alliance defendants, I would testify as follows:

I have carefully examined the performance of the regulatory processes in nuclear energy and conclude that these processes do not work and do not provide any protection to the public from injury by nuclear energy.

It is my opinion that the operation of the Trojan Nuclear Power Plant or any other nuclear power plant creates an immediate hazard to members of the public as a direct result of its creation of artificial radionuclides, such nuclides creating the hazard of cancer, leukemia, and genetic injury to the public.

It is a fallacy to think that an accident is required to create the hazard. The hazard is created the moment the radionuclides are generated in the nuclear power plant. This is so for the following reasons, my reasons being extensively supported in the research papers appended as Exhibits 2, 3, 4, 5, 6, and 7.

Reason 1:

There is no such thing as a safe dose of radiation with respect to cancer, leukemia, or genetic-mutation injury.

Reason 2:

All authoritative bodies have held that we must operate on the basis that there will be such injuries in proportion to the accumulated dose of radiation, down to the lowest doses.

Reason 3:

It is not credible that the entire nuclear fuel cycle can ever contain the radionuclides perfectly, with or without accidents. Indeed, such nuclides are released during so-called normal operations. Therefore, it follows that injury to humans is guaranteed the moment that a plant starts to operate and to create the nuclides. In fact, the killing starts even *before* a plant operates because the mining and milling operations (to obtain the uranium fuel) create massive quantities of radioactive tailing-wastes; to fuel a single nuclear plant for a single year can create a half-billion pounds of tailings. The tailings release radon gas to the biosphere—gas which would otherwise not reach the biosphere. Such radon gas and its daughter radionuclides are killing people now, and will continue to do so for at least the next billion years.

Reason 4:

The workers in the nuclear power plants receive a dose which will provoke genetic injury, and because of intermarriage with non-workers, this will result in the genetic degradation of the population-at-large, one of the most serious of all types of human injury. Although the number of regular workers at a nuclear facility may be small, the number to be irradiated by this industry is *not* small because temporary "non-nuclear" workers—even utility truck-drivers—are brought in to do the "hot" repair jobs, and to share the radiation exposure with the regular workers. Since workers start receiving irradiation the moment the plant begins operation, the injury is, in effect, established the moment the plant starts to operate.

Reason 5:

There has been gross deception of the public and public misunderstanding concerning the so-called "permissible" or "tolerance" dose of radiation. The public has been misled into believing that such doses are without medical effect, when in truth such "permissible" doses represent nothing other than a legalized permit to commit random murder upon members of the population.

Reason 6:

Even though the injury *manifests* itself after periods measured in years, the actual injury is done to the genetic materials, namely genes and chromosomes, immediately upon irradiation. Thus, it would be totally false to assume there is no immediate injury involved. The injury is immediate, is a danger now, even though visibly manifest at a later time.

Reason 7:

It is only the simplest of logic which is required to demonstrate that the essence of protection of one's health and life, and that of one's children (and their children) must necessarily reside in preventing the production of the radionuclides, since once produced, these radionuclides will guarantee human injury and deaths. The only way to prevent the production of the radionuclides is not to have nuclear power plants operate.

Reason 8:

It may be inappropriately assumed that the operation of a nuclear power plant is not an "immediate" threat to health and life. For the reasons outlined above, the threat is immediate. A simple, and highly relevant, analogy is provided by the occurrence of a fire. We do not consider it rational for one to wait to try putting out a fire simply because the flames have not started to burn our clothing or our skin. Also, we do not consider fire-fighters to be destroying property when they must hack away at furniture and other property-objects and real estate in the effort to control the blaze. Properly in a fire, we consider the threat immediate no matter how far the flames have spread at a given moment, and we take action on this basis. The situation is no different for a nuclear power plant. Prevention of the injury and death of members of the public from the operation of a nuclear power facility is a public service. I am aware of no instance in the civilian economy where we take it as a premise that injury and murder of members of the public are to be regarded as beneficent acts.

Reason 9:

Since the regulatory processes do not work to protect the public, and since the regulatory authorities continue to grant licenses for the random murder of members of the public through the licensing of nuclear power facilities, it is abundantly clear that the public can count upon no protection against victimization through the regulatory process.

□ □ □ □ □

Exhibit 1

CURRICULUM VITAE
John W. Gofman

Birth: September 21, 1918 in Cleveland, Ohio.

Education:

Grade and high school in Cleveland, Ohio.
A.B. in Chemistry from Oberlin College, 1939.
Ph.D. in Nuclear/Physical Chemistry from the University of California at Berkeley, 1943.
Dissertation:
The discovery of Pa-232, U-232, Pa-233, and U-233.
The slow and fast neutron fissionability of U-233.
The discovery of the 4n + 1 radioactive series.
M.D. from the School of Medicine, University of California at San Francisco, 1946.
Internship in internal medicine at the University of California Hospital, San Francisco, 1946-1947.

'If we listened to warnings from every environmentalist dingbat, we'd never get anything built'

Positions:

Academic appointment in 1947 in the Division of Medical Physics, Department of Physics, University of California at Berkeley. Advancement in 1954 to the full professorship, a position held to the present time, with shift to Emeritus status in December, 1973.

Concurrent appointment since 1947 as either instructor or lecturer in Medicine in the Department of Medicine, University of California, San Francisco.

Additional appointments held:

(1) Associate Director, Lawrence Livermore Laboratory, 1963-1969. Resigned this post to return to full-time teaching and research. Remained as Research Associate at Lawrence Laboratory through February, 1973.

(2) Founder and first Director of the Biomedical Research Division of the Lawrence Livermore Laboratory, 1963-1965. This work was done at the request of the Atomic Energy Commission for the purpose of establishing a program of overall evaluation of the effects of all types of nuclear energy activities upon man and the biosphere.

(3) Member of the U.S. government's Advisory Board for NERVA (Nuclear Engine Rocket Vehicle Application), approximately 1963-1966.

(4) Member of the Reactor Safeguard Committee, University of California, Berkeley, approximately 1955-1960.

(5) Group Leader with the Plutonium Project (for the Manhattan Project) at the University of California, Berkeley, 1941-1943.

(6) Physician in Radioisotope Therapy, Donner Clinic, University of California, Berkeley, 1947-1951.

(7) Consultant to the Research Division of the Lederle Laboratories, American Cyanamid Corporation, 1952-1955.

(8) Medical Director, Lawrence Radiation Laboratory (Livermore), 1954-1957.

(9) Consultant to the Research Division of the Riker Laboratories, approximately 1962-1966.

(10) Medical consultant to the Aerojet-General Nucleonics Corporation, with special emphasis on the hazards of ionizing radiation, for approximately eight years during the 1960's.

(11) Scientific consultant to Vida Medical Systems, 1970-1974; co-invented the VIDA heart monitor, a pocket-worn computer to detect and announce the occurrence of serious cardiac irregularities; invented a skin cardiographic electrode now widely used throughout the USA.

Honors and awards:

(1) Gold-headed Cane Award, University of California Medical School, 1946, presented to the graduating senior who most fully personifies the qualities of a "true physician".

(2) Modern Medicine Award, 1954, for outstanding contributions to heart disease research.

(3) The Lyman Duff Lectureship Award of the American Heart Association in 1965, for research in atherosclerosis and coronary heart disease.

(4) The Stouffer Prize (shared) 1972, for outstanding contributions to research in arteriosclerosis.

(5) American College of Cardiology, 1974; selection as one of twenty-five leading researchers in cardiology of the past quarter-century.

Publications:

Approximately 150 scientific publications in the following fields:

(1) Lipoproteins, atherosclerosis, and coronary heart disease.
(2) Ultracentrifugal discovery and analysis of the serum lipoproteins.
(3) Characterization of familial lipoprotein disorders.
(4) The determination of trace elements by X-ray spectrochemical analysis.
(5) The relationship of human chromosomes to cancer.
(6) The biological and medical effects of ionizing radiation, with particular reference to cancer, leukemia, and genetic diseases.
(7) The lung-cancer hazard of plutonium.
(8) Problems associated with nuclear power production.

Books:

(1) What We Do Know about Heart Attacks.
(2) Dietary Prevention and Treatment of Heart Attacks (with Alex V. Nichols and E. Virginia Dobbin).
(3) Coronary Heart Disease.
(4) Population Control through Nuclear Pollution (with Arthur Tamplin).
(5) Poisoned Power: The Case against Nuclear Power Plants (with Arthur Tamplin).
(6) Advances in Biological and Medical Physics (co-editor with John H. Lawrence and Thomas Hayes; a multi-volume series).
(7) Contributor of chapters to numerous books, including some on nuclear engineering, cancer induction, biochemical and biophysical research methods, heart disease, and effects of radiation.

Teaching:

(1) Application of radioactive tracers to chemical, biological, and medical problems.
(2) The biological and medical effects of ionizing radiation.
(3) Mechanisms of cancer-induction.
(4) Atherosclerosis and heart disease.
(5) Environmental factors in the induction of cancer.
(6) Epidemiological approaches in cancer and heart-disease research.
(7) Research guidance of some 25 students toward the doctorate in biophysics or medical physics.

Patents:

(1) The slow and fast neutron fissionability of uranium-233, with its application to production of nuclear power or nuclear weapons.
(2) The sodium uranyl acetate process for the separation of plutonium from uranium and fission products from irradiated fuel.
(3) The columbium oxide process for the separation of plutonium from uranium and fission products from irradiated fuel.

Current work:

(1) Continuation of research on induction of cancer and leukemia by ionizing radiations from radionuclides and X-ray sources, as well as the biological hazards of plutonium.
(2) Guidance of Ph.D. research dissertations of students in the biophysics program at the University of California.
(3) Independent consulting.
(4) Chairman, The Committee for Nuclear Responsibility (uncompensated public-interest work).

□ □ □ □ □

THE SHEEP AND I:

Some Thoughts on Alternatives to "Civil Disobedience"

I am frequently asked, "But what can we *do* to stop nuclear power?" Not all opponents of nuclear power believe in jumping fences, occupations, or blockades at nuclear facilities. Two ideas for other actions come to mind: both could provide mechanisms for educating more people about the deadliness of nuclear power, for raising awareness about the inalienable right to life, and for manifesting a fundamental principle of morality: *Individuals must be held responsible for what they do and help to do . . .* the principle briefly advocated by this country at the Nuremberg Trials.

Mental Anguish

There simply can be no doubt that one of the most serious forms of injury to humans is the induction of mental anguish. Anguish is a legally recognized form of personal injury. It comes up regularly in divorce cases as "mental cruelty".

Humans in general are horrified to think that their action may have caused the death or injury of other humans. Without a doubt, this concern is so well developed in many humans that they will suffer a severe state of mental anguish from knowing that their actions have helped to kill other humans.

Recently, the electric utility industry and its officials have taken willful action which induces mental anguish of the most severe form in a large number of utility customers. Millions of persons who need electricity in their homes and places of business are forced, by the monopoly arrangement of electric supply, to purchase their electricity from companies which have gone into the human murder-business by constructing and operating nuclear power plants, for which the customers must pay.

That nuclear power plants *are* committing premeditated random murder is undeniable. Even the Environmental Protection Agency acknowledged in 1975 that nuclear power will kill hundreds of members of the public every year, even if everything goes perfectly! (1). While the EPA has grossly underestimated the number of victims, the Agency has effectively confirmed that premeditated random murder will occur.

In 1978, a major advance in government honesty was made when the Nuclear Regulatory Commission

\+ + + + + + + + + + + + + +

(1) U.S. Environmental Protection Agency. *Draft Environmental Statement for a Proposed Rule-Making Action Concerning Environmental Radiation Protection Requirements for Normal Operations of Activities in the Uranium Fuel Cycle.* Office of Radiation Programs, EPA, Washington D.C. 20460, May 1975.

finally admitted there is *no* known safe dose of ionizing radiation, no "threshold" (2).

Thus, people who need electricity have been forced against their own desires and convictions to participate in the process of murdering innocent humans, toward whom they bear no malice. And this has induced a severe state of mental anguish in countless thousands—perhaps millions—of such persons. Indeed, in the hope of relieving the mental anguish, many have even sacrificed high earning capability in order to be active in the anti-nuclear movement.

Possible Remedies

There are at least two possible remedies to compensate for the grievous injury of mental anguish:
a.) A swarm of citizen *lawsuits* against utilities and their directors for mental anguish and punitive damages (3);
b.) A mass *direct-action* campaign against utilities to obtain at least partial compensation for the anguish.

\+ + + + + + + + + + + + + + +

(2) Confirmatory documentation:
U.S. Nuclear Regulatory Commission. *July 31, 1978, SECY-78-415, Policy Session Item from Robert B. Minogue, Director, Office of Standards Development, to the Commissioners.* Subject: Further Actions to Control Risks Associated with Occupational Radiation Exposures in NRC-licensed Activities. See especially page 11 of its "Enclosure B", in which the Office of Standards Development urges discontinuation of the term "permissible dose" because it has been *mis*interpreted to mean "safe", when in fact, "Considerations of the linear hypothesis indicate that some risk is associated with any dose of radiation, however small."

It is a pleasure to give credit where credit is due. In a fine letter to Dr. Gofman, September 11, 1978, Drs. Minogue and Karl Goller (of the same office) conclude that, ". . . tough issues have to be faced in dealing with non-threshold pollutants; we think there is growing awareness that radiation is only one of these."

(3) It may be pertinent that on August 11, 1978, the famous attorney, Melvin Belli, filed a lawsuit against the Bank of America on behalf of a client who could not get back $20,000 deposited in the Bank's Saigon branch shortly before the North Vietnamese took over the city. In addition to claiming the $20,000, Belli is suing the Bank for $1,000,000 for "emotional distress" (plus $3,000,000 additional for punitive damages).

Suppose the victims suffering from mental anguish were to get together and decide that an appropriate compensation for their suffering would be $100 per month. If they live in that "service region" on the average for five years, the amount would be $6,000—a very modest sum for such a grievous injury.

Suppose, then, that these electric customers agree to start sending bills to the utility for their injury. Thus someone receiving a bill for $40 from the utility, instead of paying the bill, simply sends the utility a bill for $60—the difference between what the utility owes him/her, and what he/she owes the utility.

In no way would this action deny the justice of paying for electricity. Rather, this action is simply a recognition that individuals have rights, not just utilities, and that there is a *two-way* transaction going on. The utility is selling power, and I am buying it, but the supplier (there is no choice of supplier) is involving me also in the murder-business, which changes the terms of the transaction.

Now it could happen that some utilities might be so unreasonable as to deny they owe *any* mental anguish payments, and they might even act rashly to threaten disconnecting delivery of electric power to the victims, even though this would be manifestly unfair.

In some states (e.g., New Hampshire), power customers have a right to a conference with the utility at the utility's office, if they think termination of service would be

Peg Averill. Art for People, Washington, D.C.

unjustified. And grossly manhandled customers who are not satisfied with the results of that conference may request a conference about the matter with a staff member of the Public Utilities Commission prior to termination of service. (Details from the Granite State Alliance, 83 Hanover St., Manchester, NH 03101).

When and if utility companies appear deaf in hearing the just claims of their customers, involvement by an increased number of customers could materially assist the utility in overcoming its hearing defect.

Visible Mourning

It is, of course, difficult to suggest simple and adequate relief of the great mental anguish being suffered by people who are forced against their wills to participate in the process of randomly murdering other humans in this and future generations. One small, but perhaps important, measure which could help provide some relief would be a *public expression of the mourning* felt for those being murdered by nuclear power.

Since the mental anguish is a very deep ethical matter, it is essential that the mourning be at a level of dignity commensurate with the deep seriousness of the problem. It should be quiet, dignified, but unmistakable to the community. Historically, a highly dignified method of expressing grief for the loss of another human has been the wearing of a black arm band by those in mourning.

A simple, black armband would do very well. However, it would be helpful for the community to know the source of the sadness, so it would be appropriate for the armband to identify the utility company engaged in nuclear murder in each specific locality where the armbands are being worn. Surely it is important to convey to those conducting the random murder, the feelings of those suffering the mental anguish therefrom. Vigils near the homes of utility directors and officers would increase their awareness.

If we are asked why we mourn nuclear power victims without also mourning fossil-fuel victims, we would have the opportunity to educate people *also* about the victims of fossil-fuels, plastics, lead-poisoning, Kepone, asbestos. . . .

Some of those older killers are intricately entangled now in our way of life, whereas nuclear power is not. In fact, all the nukes in the U.S., combined, still contribute just 3½% of our total energy supply.

The fact that some older technologies are turning people into basket-cases and corpses, is one of the strongest arguments for the good sense of *preventing* another tragedy from radioactive poisons. Visible mourning might help, as much as or more than fence-jumping.

□ □ □ □ □

7.

ABOMINATIONS UNLIMITED:

From Here to Eternity?

Most of you here today have reached the conclusion that nuclear power is an abomination which should be stopped, worldwide, at the earliest possible moment.

And beyond that conclusion, I suspect that most of you have wondered at one time or another, as I have, why it is that we have to work so hard *just to get back to the starting line* in the matter of enjoying our brief lives on this Earth. There are few phenomena which are simple, and the ideas I shall explore with you do *not* represent total answers. But there are two points which I think are exceptionally important, because many of our absurd problems like nuclear power and nuclear armaments, are very intimately tied to these points.

The first point is that man's history is the story of power and privilege wielded by *some* groups of humans over *other* humans. The second point is that *technology* has been the ever-present handmaiden of those who wield the power and have the privilege. Technology (and its science underpinning) prides itself on its neutrality and "objectivity" . . . in its total subservience to the wielders of power and privilege!

Lest the first point be doubted, it is worth noting that so ingrained is the idea that power and privilege are the *proper* state of affairs, that there are some backward areas of the world where royal families still exist and where people are shocked at the idea of giving up their kings and emperors. And although kings and emperors have rather steadily become fewer in number, earthwide, the phonomena of power and privilege have not disappeared; indeed, they may have become more entrenched, dangerous, and tyrannical.

While ex-king Farouk of Egypt may have been correct in saying that soon the only kings left will be those of diamonds, hearts, clubs, and spades, the truth is that what has replaced them and will be replacing them (barring some major societal changes) are tyrannies of *many* forms, including the tyrannies of fascist dictators, the tyrannies of communist dictators and bureaucracies, and the tyranny of the so-called democratic state.

There is little reason for the remaining utopian socialists to assume optimistically that changing the economic system is going to bring freedom from tyranny, what with the examples of the Soviet Union, China, Vietnam, Cambodia. These socialist examples don't give us much reason to believe that tyranny rests upon any particular economic organization of society. Power and privilege are very much alive-and-well in those so-called socialist countries. And, incidentally, nuclear power is avidly promoted in the Soviet Union, where, ostensibly, capitalist "greed" is not supposed to exist.

A Prime Use of Technology

From the earliest time, it became evident that if power and privilege were to be served, every clever insight which could multiply the power which one man could wield over another was most welcome—from the rock, the machine, the bow and arrow, the gun and gun-powder, up through the most elegant monstrosity, the maneuverable nuclear missile. Such clever insights, and the capability of transforming those insights into operational hardware, spawned the development of a special class of humans: the technologists.

Because of their obvious utility to privilege and power, the technologists were given a somewhat greater share of the material comforts than the bulk of ordinary humans. The technologists have served the privilege-power elite very well through history and over the globe. Not only have they provided the means for massively overwhelming an individual's physical ability to protect himself from depredation, but early on, they realized the power to be derived from mental and spiritual control as well. The technicians have provided remarkable weapons of control through the devices of propaganda, of surveillance, and of other dismal forms of mind and behavioral control.

The power-and-privilege elites have no psychological problem to overcome. They just know they are the anointed. Even those relatively *recently* anointed, like the ruling class of the Soviet Union, know intuitively that they are better than the peons. Privilege does not necessarily require that one sport the clothing advertised in *The New Yorker*. A Mao or Stalin jacket can be even more a symbol than the latest Paris creations.

The technical servant of the privilege elite *does* have a slight problem, however. He or she is commonly of peon origins and commonly of fairly high intelligence (at least with respect to handling mathematics, physics, or chemistry), so there is always the danger that he will suffer psychological difficulties in serving anti-human goals as he carries through the plans and programs of his privileged masters, be they commissars or capitalists.

Mental Acrobatics, Moral Abdication

To circumvent this problem, solutions have been found. The first is the technologists' phenomenal discovery that science and technology are neutral, objective, and thus obviously beyond reproach with respect to malevolent intentions. The second is the credo of technical excellence combined with utter faith that for every technical problem there exists a technical "fix", if sufficient ingenuity is applied to the problem. Third is the technologists' self-imposed rule that decisions concerning the applications of technology are the province of *others*.

Thus, the firepower of "conventional" ordnance, nuclear weapons, biological warfare, poison gas, other types of chemical warfare, nuclear power plants, gas chambers for human extermination, electronic surveillance, mind control—all these have occupied the minds of technologists at one time or another in places around the globe. And most still do.

When I was an Associate Director of the Livermore Weapons Laboratory, in charge of biology and medicine, I once engaged Dr. Roger Batzel (now Director of the Laboratory) in a discussion of the folly and immorality of nuclear weapons in the solution of human problems. Roger had learned the credo of "technical excellence" very well. His reply was that the job of scientists at a weapons laboratory was to make the best nuclear weapons, and to leave to the politicians and statesmen the questions of *use* of such weapons. I doubt that the idea that a technical solution to human conflicts is absurd, has ever occurred to him to this day.

I also suggested to my colleagues among the Livermore directors that if the politicians asked the Laboratory to come up with the best design of gas chambers, for a final solution to one of our minority problems, the Laboratory and others like it would vie with each other for the contract avidly. Their answer was that I was "exaggerating". Perhaps a Livermore or Los Alamos scientist would even be able to consider his assignment to the gas-chamber-problem as a noble step forward in his humanity, since the execution rate by gas chambers is certainly not as rapid as that by thermonuclear weapons.

Technical Fixes for Nukes

Let us return to the other part of the credo—for every technical problem, there exists a technical fix.

A Technical Fix from the DAFFY TAFFY DIVISION

(Your Tax-Dollars at Work)

The item below, with caption, comes straight from the January 8, 1979 issue of *Energy Insider*, a publication of the U.S. Department of Energy.

"THE TAFFY MACHINE," above left, a sticky foam personnel barrier, was demonstrated at a two-day briefing at Sandia Laboratories to show military and civilian nuclear weapons officials how the laboratories are working on safety and security. Activated by an intruder, the sticky foam material quickly fills the protected area, leaving little room for the intruder to maneuver.

\+ + + + + + + + + + + + + +

I think all of you here are by now familiar with *why* the privilege-and-power elites of communist and capitalist societies are so in love with nuclear power. It is centralized, is not a do-it-yourself technology, and thus is an integral part of the system of *control* by the power-privilege elite.

It would be hard to dream up, on rational grounds, a more stupid way to provide energy than by creation of astronomical quantities of persistent, indestructible poisons. The quantities of such poisons are not a matter of debate; rather, they are calculable quantities directly related to the amount of energy released by nuclear fission.

Now—desirable as this centralized form of energy is to the power-privilege elite—there exists the minor complication that, unless more than 99.999% of the by-product poisons are kept out of the environment, there will be a cancer and genetic price which can make current death-rates look mild by comparison. And, sadly for the elites, a fair amount of knowledge about this problem has reached the peons, and they have become a bit restive.

Meg Simonds and Lori Kearney.
Courtesy of the Abalone Alliance, San Luis Obispo, CA.

But never fear, say the elites, the technologists who have always served us *so well* and *so faithfully* will produce a technical fix for this problem! The very idea of *not* creating the monumental quantities of radioactive poisons by GIVING UP such a delightful energy source is unthinkable, in view of the beautiful features of societal control inherent in this technology.

And the technologists have indeed exercised their ingenuity in providing the technical fix. The fix comes in many forms, not the least of which is in the form of that technology known as propaganda. Let us consider some of the major elements of the technical fix and some of the super-technologists who have provided such super fixes.

Get Expert Opinion

First, there was the redoubtable Lauriston Taylor and his National Council on Radiation Protection. Their historic contribution lay in the early discovery that the poisons were not really so poisonous.

By 1954, this remarkable organization studied the problem carefully by the technique generally known in the trade as the "thin air" technique. To those of you unfamiliar with exotic science and technology, the "thin air" technique consists of developing answers by committee when scientific answers would be unacceptable to the public. So it came to pass that in 1954, this great collection of scientific minds announced that if humans each got 36 rems of radiation per year, there would be no physical effect on them.

If nuclear power had advanced by 1954 to the level which the Atomic Energy Commission would have liked, the result of exposing the population to what the National Council said was *safe*, would have been a cancer death-rate some 20 times the 390,000 who now die annually of cancer in this country. That would be some 7,800,000 per year, contrasted with only 2,000,000 who die per year now of *all* causes put together.

To imagine that the nuclear industry would have spent money to reduce the dose delivered *below* the dose declared safe, is to overlook the evidence from other industries. So it was just pure luck which prevented this monumental health catastrophe. It happened that nuclear engineering was not far enough advanced in 1954 to install the nuclear industry at that time. If any of you are under the delusion that prudence or wisdom prevented such a disaster, you are sorely mistaken.

Poison Everyone, Everywhere . . . But Just a Little

Next came the technical fix known as "dilution is the solution to pollution". This brilliant scientific development occupied many great minds, and they organized themselves into a new business known as the "health physics profession". By mixing the radioactive poisons with a large volume of air or water, they can be distributed over the whole planet, instead of just one locality. The cancers and leukemias are then harder to detect as having been caused by the radioactive poisons. What people don't know doesn't hurt them, it is reasoned. This approach is still very much alive and well, and is a major pillar of the nuclear industry.

"Fixes" Using Homes and Hearts . . .

More recently, we have had new genius applied to the problem of technical fixes. A prominent biologist by the name of Dixy Lee Ray discovered this one. It is called the "re-definition" fix. Dr. Dixy has advocated that the poisons be re-named as a resource and brought right into our economy and our homes, where we can all share in the guardianship of the radioactive waste at the same time that we use it for beneficent purposes.

Even though *she* is a Democrat politically, no less prominent a Republican than Ronald Reagan has recently been carrying the banner of the "resource" designation for radioactive waste. He sees growth for the gross national product through use of the radioactive waste in artificial hearts, metering devices, medicine, industrial catalysis, ionization smoke detectors, food irradiation, and so forth. Here a nuclide, there a nuclide; everywhere, an active nuclide. Says Ronnie (April, 1978): "It sounds as if there is a new frontier out there—right in those piles of nuclear 'waste'." We just need to develop the national will, right? A collective madness for putting poisons everyplace.

. . . And Our Stomachs

Another one of the recent additions to the technical fix armamentarium is the "store-it-then-eat-it" school of handling the radioactive waste problem. Dr. Bernard Cohen is clearly the leader of this school. He wrote a whole scholarly treatise showing that *if* you could safely store the waste for 600 years, then you could EAT up to a half-pound of it and have only a fifty-fifty chance of developing cancer. We may yet see the marriage of two, great, technological advances if we are lucky. Should human cloning come along well enough, one of your cloned brothers may even be around to eat the radioactive garbage we produce today.

A "Fix" for Bodies

Dr. Alvin Weinberg, former director of the Oak

Ridge National Laboratory, is one of the senior proponents of the technical fix approach. Yes, Dr. Weinberg *is* aware that the mishandling of radioactive poisons "could spell disaster" (*Nuclear News*, December 1971). But, says Dr. Weinberg, "Let us suppose that over the span of time I envisage, we shall learn how to cure radiation-induced disease —solid cancers, leukemias, for example—and to identify and control genetic defects induced by radiation" (*Nuclear Industrial News*, August 1974).

Well! If you can't fix the *nukes*, then fix the *people* . . . maybe. As a result of my 32 years in medical research, I consider it reckless and irresponsible to assume there will ever be a true *cure* for cancer. As for *treatment* of cancer, ask anyone who has *had* some if it's pleasant. Then deal, if you can, with mentalities like the body-fixers.

The "Safe-Handling" Fix

Dr. Jim Liverman, an acting assistant secretary of the Department of Energy, has a fix known as the "safe handling" theory. He has publicly demonstrated its effectiveness in making the hazard of radiation go away.

Recently, Drs. Thomas Mancuso, Alice Stewart, and George Kneale presented the evidence that the Hanford nuclear workers were dying of cancer induced by radiation well *below* the exposures falsely advertised as safe by the standard-setting bodies. A classic case requiring the "safe handling" technique of Dr. Liverman!

Here is how it works—the beauty of simplicity. Dr. Liverman simply cut off the research funds from Dr. Mancuso, and put the entire project at Oak Ridge National Laboratory where, as Dr. Karl Morgan pointed out, the Department of Energy can get answers more to its tastes.

If there is *one thing* which the 20 laboratories of biology and medicine sponsored by the Department of Energy have demonstrated, it is that they can be relied upon for "safe handling" of embarrassing data. *Outside* that charmed circle of DOE-supported laboratories, all kinds of scientists find that radiation is a serious carcinogen, but all 20 DOE laboratories put together never have a word to say about this subject. The few scientists who were in that charmed circle and got the "wrong" answers, are no longer in the charmed circle, where tax-dollars produce the lowest yield of useful health information known to man.

[However, the devastating weight of ever-accumulating evidence about low-dose radiation has finally forced some employees of the *Nuclear Regulatory Commission* into spasms of brave and risky intellectual honesty. In June 1978, Robert B. Minogue, Director of the NRC's Office of Standards Development, said, "The NRC's position now is that all radiation is bad, no matter how little. . . . There definitely is some cancer incidence even at the low levels of exposure (of the Hanford workers) . . . What we have found is, by God, there ain't no threshold. There are some diehards who still believe in it, but it's a myth that there is a threshold."]

The Ultimate Fix

Lastly, I must comment on the greatest technical fix of all. A couple of years ago, I did some research on the lung-cancer hazard of plutonium, and I concluded that even some of the more pessimistic estimates of hazard might have *under*stated the problem seriously. I happened to be traveling to Buffalo and happened to meet Dr. Lynn Draper, an eminent and capable nuclear engineer, in an airport. I had, by then, concluded that in a well developed plutonium economy—the sort always described by the nuclear power advocates—the amounts of plutonium handled would be so high that containment of all *but one part in a million* would be required to avert a lung-cancer epidemic.

I asked Dr. Draper, "At what level of plutonium toxicity would you say that we should *give up* the idea of a plutonium economy?" His answer was that there would be *no* such level. If the engineering requirements were to contain plutonium to one part in a billion, or one part in ten billion, the engineers would simply do it!

This is the supreme technical fix. There can never be a problem, because all technical problems will have a technical solution . . . even if it strains human credulity beyond all reasonable bounds to believe such engineering perfection is even remotely likely to be achieved in a world of fallible men and fallible machines.

My best estimate—to borrow from that great odds-maker, Professor Norman Rasmussen—is that the chance that Dr. Lynn Draper (or any combination of Lynn Drapers) can achieve the plutonium containment required, is about the same as the chance of a meteor striking you.

Keep in mind that it took seven months before an engineer noticed that a 420-ton reactor vessel, standing three or four stories tall, had been installed BACKWARDS at the San Onofre nuclear power plant. The difficulty, explained the project manager, is that the reactor and its headpiece are symmetrical. One wonders if our nuclear wizards can tell their butts from first base. The "fix" at San Onofre is choice: the headpiece may be placed on backwards too, and according to the project manager, "We'll just load the fuel from the other end." Naturally, the computer programs will

From Ireland.
By Martyn Turner in *Nuclear Ireland* (by M. Hussey and C. Craig, Irish Writers' Co-op, 50 Merrion Sq., Dublin 2).
Originally from the *Irish Times*.

* The Titanic was built in Belfast.

also be changed, "so the fuel rods will be in proper position" (*Washington Post*, November 24, 1977). *This* is the profession which promises to contain radioactive poisons with virtual perfection. *This* is the profession governed by Murphy's Law: "If anything *can* go wrong, it *will* go wrong."

Nevertheless, the power-privilege elite and its technologists have concluded that there simply is no problem with nuclear power, and they are going ahead to produce the radioactive garbage, come hell or high water. And since their wisdom comes from on high, the proper role of the peons is to accept that wisdom—an objective, scientifically reasoned wisdom, especially since it is based upon that marvelous, reliable material known as thin air.

Having a Say . . . . . . By Adjusting the Pay

We might consider the case of a utility company which insists on the privilege of random murder via nuclear power. If enough people become educated about it, then enough of them might simply end up not paying their utility bills, because they refuse to pay for random murder of themselves, their neighbors, and members of future generations.

This is not a radical idea in the usual sense, for it has just been embraced in California in a slightly different context, even by arch-conservatives. As you know, California has just passed Proposition 13, which cuts property taxes severely both for big business and for the average homeowner, and requires a two-thirds vote of the legislature instead of a mere majority in order to increase taxes of *any* sort.

Many pundits analyzed the vote as a demand just for property tax relief. I don't agree. I think people were saying they want *less government*, and are balking at having to work until May of every calendar year just to cover their taxes. Forced labor used to be called *slavery*. Involuntary service to support various government programs one would not support voluntarily is nothing *but* slavery. The income tax does not exist among *free* people; the federal income tax did not exist in *this* country until 1913. I would support its repeal. Naturally, I supported Proposition 13 as a small step in the right direction, and I am delighted that it passed.

The politicians attempted to frighten the public before the election by saying they would abolish health services, close schools and libraries, cut police and fire protection, close museums and parks. Since the election, some of the local governments have begun doing exactly those things—cutting at the bottom instead of at the top!

Paul Gann, a co-author of Proposition 13, has now publicly stated that if politicians continue responding to the vote by cutting the popular services, he is going to recommend that people stop paying their taxes altogether! And Mr. Gann is by no means a radical. So it is possible to have people come to the point where they say they have *had enough* of abuse.

Technology "X": Accept or Reject?

Those of us here are resisting abuse by nuclear power. There are additional technologies which also perpetrate random murder. So some people have turned against *all* technology, but I am not one of them.

Quite obviously, the vast majority of us like many technological advances. For example: sanitation technology and its effect in eliminating so many serious diseases; analgesics reducing the pain of surgery; the automobile; radio and even television. I can think of many additional technologic advances which I think are fine. Besides, I, for one, think science is exciting and fun.

Just *what* is it about science and technology which gets us into serious problems?

Nuclear power and nuclear weapons are so obscene that they appear to speak for themselves as undesirable excesses and monsters of technology. Still, I don't think that is good enough. We really have to have some criteria about technologies, old and new, which determine their acceptability. Otherwise, it is quite likely the obscenity of nuclear power and nuclear weapons may yet appear mild, compared to some of the *future* monstrosities which are possible. We really need to do some clear thinking about this issue.

Inalienable Right to Life?

One of our cherished ideas is that we all have the right to life as an *inalienable* right. We do not condone, at least overtly, individual murders, and we held the Nuremberg trials to punish mass murder. Nuclear power will obviously cause random murders; it is right *now* causing such murders.

It sounds good to say we are dedicated to the life principle, and so we condone *zero* random murders. The nuclear advocates counter this with the statement that, since many other technologies cause random murder, why can't they have *their* share of random murders? They do make a point.

For example, we permit the auto industry to operate using lead in batteries and gasoline, and asbestos in brake linings. We know very well that both of these poisons are getting into the environment, and they are causing murders. We do not stop the auto industry. We do not say, "Shut it down."

For example, we all know that the permitted emission levels for a whole host of chemicals is not zero, and that poisoning is occurring at the workplace and outside it. We don't shut all that down either.

For example, we provide the arms and training of military battalions for dictators around the world, for the clear purpose of murdering members of their own populations. Each of us who pays taxes aids and abets this effort, right up to today. Each of us is guilty of those random murders.

The superficial approach to these moral dilemmas is the popular "risk-benefit" analysis. Suppose we consider the logic of this dubious concept. Suppose we say that for the good of us all, *some* random murder by radioactivity would be acceptable. Just *how many* murders are acceptable, and for what benefit?

Some have proposed the formula that an activity which adds a certain number of dollars to the GNP is worth one death. Are you ready to draw the line at one death per million dollars spent, ten deaths per million dollars, 100 deaths per million dollars? Just what would be repulsive in this type of analysis?

There may be a certain macabre merit in the dollar approach, however, if you make two assumptions. First, you must assume that what people pay for something (which shows up in aggregate as its share of the GNP) approximately reflects the importance which the thing has for people in that society. Thus food, housing, private transportation, medical care, entertainment, are activities which account for big chunks of our GNP. Second, you must assume that the importance of the activity to society is pertinent to a determination of where justice lies.

One view of justice would hold that the inalienable right to life would preclude acceptance of any technology which murders people, no matter how popular the technology might be. Any compromise of this position means that the right to life is *not* inalienable—means that premeditated random murder is permitted if the majority approves.

But another view of justice might say that the Golden Rule is more important than an absolute right to life. Potential victims of a popular technology might well accept a certain chance of dying from that technology, because they do not wish to give up some *other* activity which likewise imposes a risk on others. The question for them might be, "Are my wishes imposing a risk on others greater than the risk I would be willing to have others impose on me?" In such a framework, knowing the *size* of the risks involved is fundamental to justice, and the greatest betrayal of one's

fellow man would be to suppress or obscure information about the magnitude of the risk. Dishonest handling of such information has characterized the nuclear controversy from day number one.

The Golden Rule provides a slippery framework, especially when it comes to our responsibility to future generations. Also it leaves a lot to be desired unless there is a way of considering *aggregate* risks to self and society. A person who thoughtlessly answers that one chance in one thousand per year of dying from nuclear power is OK for himself, might well object to killing 200,000 fellow Americans per year . . . which is precisely what such odds really mean in a population of 200 million people. And what if 1,000 different pollutants *each* produce "only" one chance in one thousand of killing you per year? You'll be gone in a year, that's what.

Still others stress that what must be inalienable is the individual's right to *choose* whether or not he wishes to participate in a random benefit-risk exchange. Freedom to

AUTH
'Nuclear waste problem? I don't know about you, but I don't
want my kids growing up in a world where there aren't any problems left to solve!'

choose is the paramount right, they assert. This means that society could engage in almost no activities at all, of course, for who can prove that the one person who chooses to opt *out* of participating in an activity (for instance, flying), won't become a victim (when a plane crashes into a street or highway, for instance)?

Or suppose individuals freely *choose* the risks of working in a nuclear power plant. As a result of the genetic damage which they are assuredly going to sustain in such work, the children they conceive carry some defective genes which will make life miserable or brief for some humans in even a *later* generation. Those unfortunate victims did *not choose* to participate.

Some give a different perspective on risk-benefit analysis, by suggesting that a technology which *saves* more lives than it causes to be *lost* is obviously an acceptable technology. Is it? What, for instance, about so-called nuclear medicine, which delivers radioactive poisons into the environment in growing quantities? Why is it supposed to be acceptable that some persons, on a random basis, are condemned to die so that some others, probably equally random, may be saved? This is also, of course, the justification commonly presented for wars. Do we approve of wars on this basis?

We need some clear thinking about people's right to damage their *own* bodies, in contrast to damaging someone *else's* body—for instance, by tolerating nuclear power. And about people's right TO LIVE, when it interferes with an activity desired by the majority! Of course, these moral questions pertain to many technologies besides nuclear power.

At the present time, society operates as if premeditated random murders by pollution *are* justified, unless there are "too many". But how many are too many? I wonder if every person here has reached a firm, private decision about how many mentally damaged children are too many in exchange for using leaded gasoline, and how many cancers are too many in exchange for using asbestos brake-linings in the very automobiles we used to get to this rally. What should be the criteria for permitting, or not permitting, a poisonous technology?

Irreversible Acts

Suppose we start at the *easy* end of the scale. We just do not accept technologies like nuclear power and nuclear arms whose indestructible radioactive by-products make total and *irreversible* poisoning of the planet into a real possibility.

Irreversible pollution is a moral issue of enormous importance. There is a new dimension to depravity when humans go beyond hurting their *contemporaries* (who have at least a chance to fight back), and start reaching into the future with technologies which necessarily compromise health for all generations to come.

Now, in view of such a sensible and powerful criterion, why do we have so much *trouble* eliminating nuclear power and nuclear weapons from the world?

The answer: power and privilege. Power and privilege have bought the scientific and engineering sycophants to tell us they will accomplish the miracle of containment. Power and privilege have bought the propaganda media and sycophants to frighten the working man into thinking that he will "starve in the dark" unless nuclear power goes forward. Of course, we know now, from abundant studies, that nuclear power is the biggest rip-off *yet*, from the point of view of standard-of-living and jobs for the working man!

But the point to remember is that power and privilege have been able to create the framework of *deception*, inside and outside of government, which leads society to violate the criterion that we should not chance the total and irreversible poisoning of the planet.

Throwaway Planet

Engelhardt in the *St. Louis Post-Dispatch.*

I mentioned earlier that the nuclear power advocates often ask why *they* are singled out and not allowed to help destroy the planet, since others are permitted to do so. They could point out that lead and asbestos can also irreversibly poison the surface of the planet where life is sustained. In this, they would be correct, for these substances are *not* bio-degradable, and their half-lives are infinite. Are asbestos and lead indispensable for the good life? I have the distinct impression that we may be able to do without them, too.

Such poisons as lead and asbestos deserve early and serious attention—but not as justification for *compounding* the irresponsibility by adding nuclear power.

Even the government (HEW) estimates that 5.6 million American workers have been condemned to die from asbestos exposure, and that's *not* counting exposure of the general public. When the Nazi's killed 6 million Jews, we properly called it a holocaust, but where is the outrage over the asbestos holocaust? It was not done innocently. Data on asbestos have been available for over 50 years.

Why is it now that the people who are trying to *prevent* a similar tragedy—only worse—from nuclear power, are called extremists, communists, and "enemies of the American dream"? The reason: the privilege-power elite wants nuclear power, and there is an ample supply of well-educated sycophants to do its dirty little propaganda jobs. In fact, when you have watched the scene as long as I have, you may conclude that no matter how dirty the job, there will be a well-educated, well-mannered, slick if not positively slimey creature, complete with credentials, out there somewhere available for hire.

Our More Fundamental Dilemma

Nuclear power and nuclear arms are not our society's only outrages. I would like to propose here that the very same forces which are giving us the nuclear twins are responsible for innumerable additional outrages too. And those forces are, of course, power-and-privilege elites operating through their multiplicity of sycophants.

Technology has given the world's elites the miracle known as plastics, out of which has been fashioned a replacement for the Golden Rule, which is the Plastic Rule:

''Screw unto others as you can assuredly count upon them screwing unto you''.

The two leading power-and-privilege elites which operate by the Plastic Rule are in the Soviet Union and the United States. Each has its coterie of hangers-on, groveling before these two out of fear, or simply for a bone thrown to them occasionally. It is these two power-and-privilege elites who give us nuclear power, who give us nuclear weapons, and who, unless checked, will assuredly give us nuclear war.

Their total violation of the cardinal principle stated earlier concerning choice of technologies, and their willingness to commit genocidal war to resolve the problems they themselves create (while at the same time hypocritically talking of SALT agreements, human rights, and liberation of peoples) are contemptible.

Unless we can learn to liberate ourselves from these privilege-and-power elites and from the Plastic Rule they live by, nuclear power is far from the last indignity they will heap upon the living species of this Earth. Real liberation, of course, means also preventing the appearance of any new, equally horrible, privilege-and-power elites.

I don't have the answer for dealing with the power-privilege elite problem, but solutions to problems come after we have identified the real problem. A little while ago I defined an entity known as ''uncivil obedience''. Whatever type of society you live in—democratic, socialist, or other—it is customary that you commit ''uncivil obedience'', since that is the only behavior recommended by the power-and-privilege elites.

''Uncivil obedience'' is defined as that wonderful state wherein you declare yourself to be a moral cripple, a *de facto* slave contributing your labor and dollars to a variety of outrageous offenses against humanity—nuclear power being only one of them, and not at all necessarily the worst which will be conjured up for you. ''Uncivil obedience'' is the essence of being a so-called ''good German''; unfortunately the 98% or more of us who pay our taxes to support activities with planned and built-in large-scale random murders, activities such as nuclear power, are indeed "good-Germans". We really are not different from the Germans who sat by and accepted, or who abetted, the holocaust.

It is obvious that the solution is *not* simple. The Germans who sat by or who participated in the atrocities committed by the Nazis knew that the Nazis had the guns, meaning the power, and that the Nazis had not the slightest compunction about using them against anyone who refused to comply with "uncivil obedience". Large segments of the world are clearly already in a new Dark Age, where seeking a solution is very dangerous. Thus Russians who make even the mildest noises about non-compliance with the outrages going on there are either sent to psychiatric institutions for rehabilitation or the slave labor camps.

If there is any hopeful place where change can come, I think it is in the United States, where, despite all the depredations of the power-and-privilege elite, there still exists the opportunity for expression and communication of ideas. And that is truly very important.

The Only Game in Town

The optimistic point is that large numbers of persons in the United States, and everywhere in the world for that matter, would far prefer the Golden Rule to the Plastic Rule. Every effort needs to be made to increase those numbers, even if today they represent a small minority. Every single addition to those ranks is important. I sense that, in spite of all the disillusionments and the hopelessness before the guns of power, more people are coming to realize where the problem lies, and are more ready than ever to do something about it. Even the mini-example in California, of people telling the power-structure that they think they have had enough, should not be taken lightly.

The nuclear power and nuclear arms issues are excellent tools to help teach us the hypocrisy of governments both in the USA and the USSR, to teach us that governments are simply instruments of the power-privilege elite in both areas. But do not expect that power and privilege are going to yield gracefully. Let me remind you of Frederick Douglass' important and historic admonition over 100 years ago:

Courtesy of Ray Collins of the *Seattle Post-Intelligencer*.

"Find out just what people will submit to, and you have found the exact amount of injustice and wrong which will be imposed upon them."

It is not a bit different today. Power and privilege will happily use, to borrow Mao Tse-tung's phrase, "the power that grows out of the barrel of a gun" to prevent the Golden Rule from ever replacing the Plastic Rule. And do not overlook the fact that nuclear power itself can and will be used as the justification for removing as many civil liberties as possible, to prevent any threat to the ultimate ideal of robotized humans who stay in their proper place.

Nuclear power is a symptom, albeit a very serious symptom, of societal disease. But the disease is the *existence* of privilege and power. Unless we can learn to cope with that disease and to build the numbers who would like to avoid the anti-human depredations caused by that disease, there is very little to which we can look forward. It will be a tight race to win before there is not much *left* to win, but there really is no other game in town.

Two years ago, who would have predicted the profound effect of the Clamshell Alliance in getting people to understand what it is all about? There is good reason for hard work, applied ingenuity, and optimism. NO NUKES!

☐ ☐ ☐ ☐ ☐

8.

THE NUCLEAR DOUBLE-TROUBLE:

Some Approaches beyond Widget-Fixing

> Those who *know* the truth are not the same as those who *love* it.
>
> *Confucius*

This weekend is the 33rd anniversary of the introduction of the era of nuclear blackmail. The nuclear bombing of Hiroshima and Nagasaki, far from marking the end of the nightmare caused by the Nazis in Germany and the Imperialists in Japan, was the beginning of a greater nightmare overhanging the future of humanity everywhere.

Nuclear blackmail has two aspects in this country. First: if you don't participate in the nuclear arms race, you will be incinerated or enslaved by an enemy. Second: if you don't embrace a civilian nuclear power program, you will supposedly suffer an economic disaster whose central feature will be a loss of jobs.

Like all blackmail, nuclear blackmail is a bad bargain for its victims.

In exchange for a source of extremely expensive energy for which no need exists, society is forced to participate in history's greatest crapshoot with respect to health and life—for both this and future generations. In a world-casino, we create an astronomical quantity of some of the most persistent poisons imaginable, and the game we are told to play is one of hope—hope that the engineers and scientists can figure out a way to contain all the radioactive garbage. If the scientists and engineers could succeed in this gigantic containment experiment, we would get some energy which we can readily get much more cheaply from benign sources. If the scientists and engineers continue to fail in containment of radioactivity, people will get a legacy of cancer, leukemia, and genetic injury which will continue to plague mankind for generations. As Dr. Chauncey Kepford has pointed out, the radon gas problem from milling uranium extends the injuries out to *billions* of years.

In the other part of the casino, the bargain is equally bad, but it is more difficult for the public to perceive it, thanks to a myth. The myth is that the only purpose of having nuclear weapons is to deter the use of similar weapons by an enemy. The lullaby goes like this:

The deterrent will deter,
And war will not occur!

Why the Nuclear Arms Race Has to End in Holocaust

Since so many Americans appear to believe that nuclear weapons will never be used, it is important for me to state why I think it is *nearly a certainty* that they will be used in the next 10 or 20 years, and used in a big way, intentionally. Unless, of course, people like you succeed in making the crucial difference. I offer my viewpoint as a former Associate Director of Lawrence Livermore Laboratory (1963-1969) where about half of the USA's nuclear weapons are designed. But nothing I say is classified.

Governments—for instance, the USA and USSR governments—commonly point to the absence of nuclear war in the past 30 years as *proof* that nuclear weapons successfully deter nuclear war. My view is that this will remain true ONLY UNTIL one side achieves the technical breakthroughs which give it a "first strike capability"—which is always defined as the ability to initiate and win a war while suffering only acceptable damage in return. In 1941, for instance, the Japanese evidently *thought* they had a first strike capability; nations do not attack others if they expect UNacceptable damage, according to the meaning of the word "unacceptable". However, since the Japanese were mistaken about having a first strike capability, the carnage on both sides was terrible; the USA ended the war with a technical breakthrough—the atomic bomb.

Technical breakthroughs continue, of course—a fact which undermines the key assumption of deterrence, namely the assumption that it is *impossible* for either superpower to achieve a technical breakthrough which would give it a first strike capability.

Every few months, unclassified news appears about advances in anti-submarine warfare, anti-satellite weapons, lasers, particle-beam weapons, anti-ballistic and anti-cruise missile weapons. Add to this a host of technical developments on both sides which are successfully kept *secret*, and I have to view the myth of eternal deterrence as groundless, irresponsible, and reckless "wishful thinking." Furthermore, achievement of a first strike capability is not exclusively a matter of technical breakthroughs. Later, we will examine what our rulers consider to be "acceptable damage."

I know our nuclear weaponeers. Their very highest priority is to prevent the Soviet weaponeers from developing and then using a first strike capability on us. I know that some of you refuse to impute evil intentions like that to the Soviet Union, but when we see how the Soviet rulers terrorize and thereby enslave their own people, I think it is foolish to think they would not enslave us too, if they could achieve a first strike capability. I wish to make it very clear that I am not referring to the Soviet peoples; I am referring to a small group of bullies at the top.

Some of you who don't know me, may be jumping to the conclusion that I support nuclear weapons.

Nonsense. Like you, I regard nuclear weapons as an abomination which mankind must get rid of. It's a question of HOW. And it's a question involving not only Soviet, American, and European destiny; three or four *billion* souls can be easily enslaved if only ONE super-power remains on Earth with nuclear weapons.

American weaponeers, entrusted with the defense of this country, necessarily have to assume the worst about the intentions of the other side; any other assumption in their profession would represent gross negligence. This is an important point, sometimes overlooked because of our disgust with any preparations for genocide.

Conceiving a First Strike Means Using It

Now for a few moments, please put yourself into the shoes of the American weaponeers. In order to prevent the Soviet weaponeers from developing a first strike capability, our weaponeers have to develop defenses which can *nullify* the first strike power of new Soviet weapons. And clearly, the only way our weaponeers can develop defenses against a new weapon-system is first to dream up the new system for themselves. There is simply no time to wait to *see* what the Soviets have dreamed up, now that missiles can go from continent to continent, and from submarine to land, in a matter of minutes. So our weaponeers necessarily have to dream up first strike weapons themselves, if they are going to dream up our defenses against them.

Thus regardless of the pious claims that *this* country is not interested in going for a first strike capability itself, logic tells us that figuring out a first strike capability HAS to be our very TOP priority. This is logic, not inside information. As long as competing gangs of rulers are allowed to use *force* to resolve the conflicts they create in the first place, it is simply inevitable that the parties to conflict must necessarily seek a first strike capability, under the strictest military secrecy.

Most unfortunately, the logic proceeds one step more. What would happen when our weaponeers dream up a first strike capability for which they can dream up no defense? They would simply have to assume if THEY can do it, so can their counterparts in the Soviet Union. At this point, the survival instinct would almost certainly overwhelm competing moral considerations. In an act of *self-defense*, the USA would almost surely rush to USE its first strike, as the only way to prevent a similar system being used upon us at a slightly later date.

Of course, the odds are approximately even that the *Soviet* weaponeers will be the first ones to achieve the first strike capability. Or to *think* they have. A miscalculation by one set of weaponeers just guarantees utter devastation on ALL sides instead of on one side.

If you accept my premise that technical breakthroughs in the next decade or two may indeed provide one super-power or the other with a first strike capability—and remember that such an achievement depends also on the rulers' secret standard of what constitutes ACCEPTABLE damage to their own peoples—then it follows that the intentional use of nuclear weapons is a near certainty. While experience in a nuclear weapons lab was not necessary to figure out this logic—and indeed, some others who have *not* had such experiences have reached similar conclusions—it is significant that my experience at the Livermore Lab is not able to provide me with ways to *fault* the hideous logic I have just outlined.

As I see it, no matter which side gets the winning dice in this area of the nuclear casino, the likelihood of intentional nuclear holocaust somewhere, and fairly soon, is far far greater than the American public presently realizes.

Widget-Fixing and Worse: Unilateral Disarming

If you are tracking with me this far, then it will be clear why the secret research into new weaponry by both super-powers is where the action really is. And from such

AUTH

"Don't laugh! He used to be a nuclear physicist."

analysis, it follows that it is as MEANINGLESS for concerned citizens to fuss over counting old warheads, reducing "over-kill", trimming percents off of military budgets, or ratifying new SALT agreements, as it was for concerned citizens to fuss over the improvement of a valve or filter-system on a particular nuclear power plant, instead of trying to STOP nuclear power. Neither the danger of nuclear weapons, nor the danger of nuclear power, can be reduced by widget-fixing. Therefore we *must* think big.

I believe for the United States to make *small* unilateral gestures in disarmament would be merely wid-geteering, and that for the USA to take *large* unilateral steps in disarming would be tantamount to giving the USSR a first strike capability.

Though my position on this may be presently unpopular with you, and I know that many of you are pacifists, I do oppose unilateral disarming. I believe that the amount of trust the Soviet leaders deserve is zero. And I think that accepting the idea, "Better Red than dead", would be the same kind of error the Jews in Europe made, in their refusal to believe in the total barbarity of the Nazi Germans. I even believe that the Jews in the Warsaw ghetto who took a few Nazi monsters *with* them as they were killed, are more of an inspiration for the defense of human rights than people who let monsters commit their monstrosities with impunity. I realize that such a statement must grate badly on some of you, but since I have been listening to other points of view carefully, I ask only that you give mine fair consideration. We all share a very deep desire to solve the same problem.

I would like to add one man's opinion. It is only that—one man's opinion—but the man is the great Russian dissident and "father" of the Russian hydrogen bomb, Andrei Sakharov. Recently he met with Joan Baez in Moscow, and spoke into her recorder. Asked if he is a pacifist, Sakharov replied, "I am perhaps too old to be a pacifist. I have seen too much . . . You don't understand this country. You don't understand this system, this government. Your country must stay strong."

What Difference Does It Make, Anyway?

I have been asked, in as much as I regard nuclear *holocaust* as so likely to occur, why I bother fighting nuclear power and other sources of cancer and genetic degradation. That's an important question.

"What difference does it make if they pollute the world, if they are going to blow it up anyway?" say some intelligent non-activists. My answer is that it makes *no* difference. It makes no sense to fight nuclear power unless we *also* stop the system which is going to give us nuclear war.

That fact has been troubling me profoundly since the very beginning of my involvement in the nuclear-power fight. But some of us are only ''minimal bright'', after all, and so it has taken me all this time to synthesize a concept which I think may be *helpful* and therefore worth discussing . . .

The Lust for Power As a Medical Disease

Although it is obvious that ordinary humans have only misery to gain from nuclear weapons and from nuclear power, it is equally obvious that governments globally are bent on acquiring more, not fewer, of these monstrosities. We simply must figure out *why* this is so, if we are going to stand a chance of reversing the momentum.

I have a simplistic view of the origins of the human dilemma, a sort of medical view, and some anthropologists and psychologists will disagree with it. Nevertheless, I shall share it with you as a basis for dialogue on this all-important issue.

Humans, like other animals, have the instincts of survival, which are enormously powerful instincts. Included in these survival instincts are acquisition of food, shelter, and freedom from predators. But whenever a function exists in an organism, almost invariably ERRORS also exist in the function. With the exclusion of accidents and trauma, this is what medicine is mostly about.

One major such error, and it seems to be peculiar to humans, is a *gross* distortion of the survival instinct. That distortion, pervasively evident in just about the entire recorded history of humans, leads to the desire in certain humans to acquire unlimited power over other humans and resources. Power, in this aberration, seems the only thing which makes such diseased individuals feel safe and secure about their own survival. While it appears that *most* social animals live in some sort of power hierarchies or pecking orders, the arrangement among *non*-human species does not threaten the whole species, and may even provide breeding and survival benefits for the species as a whole.

Not so among humans. Among humans, we observe cases of INSATIABLE appetites for power. We see the aberration in the form of humans who feel threatened, once they have accumulated half a billion dollars, unless they can acquire *another* half-billion dollars at the earliest possible moment. It may be that, to some individuals for whom taking control over everything on Earth is not enough, the space program is regarded as a glorious opportunity. These individuals may even sit up nights wondering if, out somewhere —sitting on the Moon and beyond—there might be some still independent ROCKS over which they could exercise their power!

The Now Society

First it's sex, then money. Then you weed out, simplify and you find power is enough.

This aberration of the survival instinct, though obviously manifest in the economic sphere, is by no means limited to greed. Nevertheless, the *visibility* of greed over the centuries of recorded time has led many concerned humans to believe that the cure for this sickness would be the elimination of economic gain—for example, the abolition of capitalism. Of course, that direction was tried in Russia, but the lust for power is very much alive and thriving in the Soviet Union, even though personal wealth is not the objective of the rulers there. Wealth is only one way of having power over others.

That the Soviet's party-faithful, their scientific elite, engineers, and athletes are given some luxuries in the USSR is quite true, but that is really no different from the doling out of modest luxury by the power-structure in the non-communist world to scientists, physicians, lawyers, the military, and even to part of the laboring class, in the effort to buy loyalty. And it works, because these servants are satisfying their own insecurities about survival. If a person's obsession with gaining power is strong enough, he usually makes the perfect, and perfectly unprincipled, sycophant. Even though he may start at the lowest rank in the structure, he will usually manage to scramble and ingratiate himself into a spot far above his starting point. In our system, such climbers may be called "self-made men" when they hit the millionaire category.

The disease, whether in a capitalist or a so-called socialist economy, is a lust for power. Since it is obvious that power-lovers seldom if ever operate in the best interests of the vast bulk of humans, they have to take care that they keep the vast bulk of humans under control. A variety of ruses are employed for this purpose. Although their objective is power and control over humans—keeping them in their proper place—the methods are disguised not merely as "the way things ARE", but also as the way things HAVE to be, as a sort of natural *law*.

The Disease Leads to Nuclear Power

Control is the main attraction making nuclear power so popular with the power-privilege class of both the Soviet Union and the Western democracies. Every aspect of society is arranged, if possible, to centralize control, to guarantee the exacting of tribute from the peons of society. The control of energy supplies is just one of the manifestations of control in general. That is why decentralized solar power gets only lip-service from our rulers.

One very successful way to control peons is economic blackmail. Do as we want, or you will lose your job. In general, it is not presented as bald-faced blackmail, but rather as the dictate of an impersonal *economy*. It is a most effective ruse, for a *normal* survival instinct is present in people at all levels. And not too many people out of the total are more than a few paychecks away from the breadline. There is no other way except economic blackmail to account for the scientists, the engineers, and the physicians, who know about the real hazards of nuclear power, for example, but do not speak out. Or, if they do speak out, they repeat such rubbish as, "A solution for managing radioactive poisons will be found," or another variant like, "A cure for cancer will be found." The panic of coping with their immediate survival requirement, which is the secure income, is so great that they construct a wall of rationalizations which prevents them from clear thinking about their own long-term welfare or that of their children, and *their* children.

As I see it, the ultimate reason that humanity finds itself witnessing a program to make people everywhere artificially dependent for energy upon some of the deadliest and most persistent poisons imaginable, is that the species has let control of society pass into the hands of its greediest, most power-obsessed, least principled members. And society lets the control *stay* there thanks to economic blackmail. When people themselves are polled about their own desires regarding energy, at least in this country, they overwhelmingly declare themselves in favor of solar power. But their desires do not count.

The imposition of nuclear power is one of the things I have in mind when I reach the following conclusion: The aberration of the survival instinct in certain humans is a most serious disease—far more serious for the species than cancer or heart disease or even the plague. For everyone here understands, I am sure, that acceptance of nuclear power means the acceptance of premeditated random murder by radiation, as a legitimate policy of modern so-called "civilization". And yet the power-elite seems ready to stop at nothing to ram nuclear power down our throats. In service to the elite, power company hirelings even photograph license plates in order to identify people attending anti-nuclear rallies—which would certainly help the police to round them up at a later date, if ordered to do so.

Resisting with Civil Disobedience

In your efforts to help stop nuclear power, some of you have decided to commit non-violent civil disobedience. Laying your body on the line is a powerful statement and educational tool, as long as it enables you to multiply the number of people who thereby come to agree with you about the importance of your stand.

Nowadays, the media will cover your story, in case there is blood spilled on the ground, but unfortunately it will carry hardly a word about the substance of your cause. At the Seabrook rally, the media people almost never left their tent, where they were apparently waiting for action of the conflict-kind. I read reams of coverage about Seabrook afterwards, and there were not even three paragraphs reporting the substance of the giant protest, though plenty of substance was publicly discussed at the rally. To make your case comprehensible to the general public, you will have to continue the tireless person-to-person education which deserves the credit for bringing the movement as far as it *has* come.

I think the power of your action depends upon the extent to which you make the case that it is NUCLEAR POWER which is violating the law. How can YOU be violating the law, when you are trying to *prevent* a crime—namely the premeditated random murder which is committed by every nuclear power plant in the country. Indeed, the random murder starts even before the plant is built, because the mining and milling of uranium *start* the murder process—politely referred to as "health effects" by government regulatory agencies.

It is clear there is going to be no relief from random murder policies under the Carter Administration. The Nuclear Regulatory Commission's most recent Seabrook decision is hardly a decision to stop nuclear power development; it may not even stop the construction at Seabrook for more than a few weeks. [*How true.*]

The fact is that long ago, the government teamed up with industry to perpetrate such a fraud and deception concerning the safety of nuclear power and the danger of ionizing radiation, that it makes Watergate look like "small potatoes" by comparison. The cover-ups have been so gross that even Peter Bradford, a newish member of the NRC, has stated (according to the Union of Concerned Scientists) that truth was one of the "first casualties" of the government's program to promote nuclear power, and that the truth has been undermined by what Bradford calls "silenced concerns, and rigged or suppressed studies". Suppression has been practiced both on engineering *and* biological considerations. Anyone within the Atomic Energy Commission who dared suggest that radiation caused cancer and leukemia was immediately excommunicated, and subjected to a determined campaign of vilification and slander. I can tell you that from personal experience. And that policy is alive and well in Jim Schlesinger's Department of Energy *today*, under the apparent guidance of Dr. James Liverman. Therefore it is right not to let the system divert much of our efforts into costly and futile hearings before *government* panels.

Challenging the "Right" to Issue Murder Permits

If we had a *justice* system in the United States, instead of a *legal* system, I doubt that there are enough jail cells to accommodate the deserving members of the atomic energy establishment, for their crimes committed against humanity. I am happy to note that the Union of Concerned Scientists has formalized the activists' call for criminal proceedings in these matters. Let me quote some recent testimony by Daniel Ford of the UCS before Congress:

> "The question of possible criminal activities on the part of officials entrusted with protecting the public safety must be resolved. As part of its background review of proposed nuclear licensing regulations, we therefore recommend that Congress request . . . the Criminal Division of the U.S. Department of Justice to

carry out an investigation of the conduct of former AEC officials, now NRC officials, to determine any role they may have had in a nuclear safety cover-up.''

That's how the legal system should be dealing with the *perpetrators*. Now, how should the system be dealing with the civil disobedients? In my opinion, the civil disobedients have to make the case that the legal system has no right to try them on the basis of criminal trespass. The sanctity of property is automatically forfeited if that property is being used to violate people's inalienable right to LIFE. Just because Congress authorized an agency to issue a license for a nuclear power plant to commit murder, does not make it lawful. Under the Nuremberg Principles, this country declared that individuals have a *duty* to consider principles which transcend obligations imposed by the state. Furthermore, there is no provision in our own Constitution empowering Congress to issue permits to murder people at random. That simply has to be challenged. *

There is a whole profession in this land known as the law profession. To be sure, most of the people in that profession serve to maintain the power structure, not to challenge its depredations. Nevertheless, there are *some* humans in that profession, and I think only a few of them are doing their job. Why are there not SCORES of lawyers in this state loudly protesting the conviction of Alliance members for criminal trespass? Where is their concern for justice? It would be nice indeed to have their support in explaining to the public WHO really ought to be on trial, and for WHAT!

\+ + + + + + + + + + + + + +

* *Honicker vs. Hendrie*, a lawsuit which challenges the ''right'' of the NRC to commit premeditated random murder by licensing nuclear power plants, was filed September 6, 1978 in federal court. Joseph Hendrie is chairman of the NRC. Jeannine Honicker is asking that the U.S. District Court in Nashville require the NRC to revoke the licenses of all nuclear fuel-cycle facilities, because murder-licenses are unconstitutional. Initial testimony for Honicker was heard on October 2, 1978. Details are available from Joel Kachinsky, attorney for the plaintiff, FARM LEGAL, 156 Drakes Lane, Summertown, TN 38483.

The Disease Leads to Nuclear Weapons . . . and Their Use

Nuclear weapons are *another* thing which justifies my earlier statement that aberration of the survival instinct in some humans is a more serious disease than cancer, and heart disease, and the plague.

In the quest for power-extension, *force* has been useful for those seeking such power-extension. The use of force creates the need for so-called "national security", and requires a willing supply of cannon-fodder. Lo and behold, people have been willing to *die* for one set of rulers, when the competing set appeared to be even worse. That was the case with many Americans during World War II, and as I have already said, some of us find ourselves thinking that the American set of rulers today is the *lesser* of two evils.

But we peons have never enjoyed a choice between war and PEACE, between domination and NO domination, between blackmail and NO blackmail. I think it is our job in the anti-nuclear movement to figure out a way finally to create this choice for mankind. If not now, when? If not us, who? I don't have the answers, and I speak here only because I think I have some of the right questions. The answers may come from *you*. Your ideas and your achievements, for instance in building alternative communication networks and alternative technologies, are formidable, and most encouraging about what is possible.

I hope we can all accelerate our thinking, because meanwhile the military and technical establishments, inside and outside universities, jointly proceed with the effort to learn how to FIGHT and to WIN nuclear wars for the power elites which they serve. With respect to achieving a first strike capability and using it, the time-schedule is closely related to their concept—voiced only through underlings, if at all—of what constitutes "acceptable damage".

When you read the writings of the military-technical establishments over the past couple of decades, you observe that they talk less and less about the human casualties of nuclear war, and more and more about how many years it would take to bring the American economy back to the 1970 level. Mr. Bassett of the Federal Preparedness Agency explained it on a recent NOVA television program about American civil defense in the event of a nuclear war. After describing how bureaucrats, and computers, and records would survive a nuclear war thanks to elaborate underground bunkers, built in nine regions of the country just for them (not for ordinary citizens), Mr. Bassett said:

> "This idea that we're going to be wiped out and nothing [more] can happen, and we might just as well give up if we have a bombing is, is, is not right, because we *can* reconstitute Federal Government. We could have a *viable* government . . . . Sure, maybe we'll be living at a lower level of economy, but we'd have enough people. Maybe we'd have to go back to a 1920 status of people and numbers and economy, maybe even earlier than that. But we've got enough people, and we've got enough people in the Federal Government who know how to do things, and who know their responsibilities, so that we could have a viable government again."

Mr. Bassett, who looked a bit overfed on the TV screen, has a wonderful view. Even by enduring a nuclear war, the public could not get government off its back. Out of the bunkers and into the rubble would crawl a healthy horde of Mr. Bassetts to tax the survivors, and to rule on who owns the viable parts of the new 1920 economy, and who owns the rubble.

Mr. Bassett's point of view should be no surprise, since his job is to please the

AUTH
'GREAT NEWS! WE'VE INFLICTED UNACCEPTABLE
DAMAGE ON THE OTHER SIDE.'

power-holders at the top who have PROVEN that they could hardly care less about human suffering. When I say proven, I have in mind abominations like arranging economies so that about a billion people are malnourished and afflicted with unnecessary diseases; like always *resisting* measures to stop the poisoning of industrial and agricultural workers; and in Russia, the enslavement of the entire population through terror.

The Task Ahead: Disease Control

For people like you, who think there may be such a thing as morality and inalienable human rights, who think the life-experience is interesting and worthwhile, eliminating the twin nuclear problems is fundamentally a struggle between humans and the disease known as power-lust . . . an aberration, a sickness, in the normal survival instinct of some humans. Whether this severe sickness is genetic in origin, I do not know. It may be. But I really don't think it matters whether it is or not. Not all humans manifest this disease, and some manifest it only mildly. Even if some fraction of every human generation will have an aberrant survival instinct, there is JUST NO REASON to allow such individuals to inflict the consequences of their disease on the rest of mankind.

Throughout history, the MOST diseased humans have been *permitted* by all the others to get to the top of the heap. Today, this means that they have the power to impose both nuclear energy and nuclear weapons on the rest of the species. To me, it is absurd and almost unbelievable that the Soviet and American populations permit their rulers to possess nuclear toys which they can freely use to poison or incinerate millions or billions of people for no reason at all.

In the face of monumental public confusion, misconception, and resignation, it is self-defeating to focus

Courtesy of Tom Darcy of *Newsday*.

THE GOOD LIFE

on widget-fixing. The public must be helped to understand that the deterrent will NOT deter nuclear war, and that nuclear energy perpetrates premeditated random *murder*. Those insights are what activate us, and they will activate others too.

The task of the activists, as I see it, is to learn how to make it IMPOSSIBLE for the sickest members of the human species to control all the others. In other words, we must eradicate the disease of power-acquisition permanently from human society. It must be gone before *new* abominable technologies are developed. For instance, mind-control. For instance, human cloning. And it must be gone before the Western and Soviet rulers get around to *merging* their power, for their greater security and consolidated control.

If the human experiment is to have a future, the only hope I can see is eradication of that disease—power-acquisition by some humans over others. It is far more important to eradicate that disease than it ever was to eradicate bubonic plague, tuberculosis, coronary heart disease, or cancer. And it is essential to eradicate power-acquisition everywhere on Earth, not just in the United States.

Some Reasons for Optimism

It is very easy to be pessimistic about all that I have just described, on the grounds that there is no reason to believe that the eradication of the power-disease is possible. After all, it has been THE way of life for centuries and millenia, everywhere on Earth. But I have recently been feeling considerably more optimistic, because of observing one human trait which we have not yet discussed in this review—namely, a desire, respect, and striving for *justice*.

It is rather amazing that some humans yearn for justice, considering that it is so rare a commodity and considering especially that a quest for justice can, in the short term, run squarely into conflict with the basic instinct for survival.

Yet the evidence is there and unmistakable that the number of humans who have placed justice highest on their list is not negligible, and this has been so over centuries. Many with deep religious convictions have for a long time felt strongly about justice, and here I refer to individual convictions rather than the generally spotty performance in this regard among organized religions. There have been the conscientious objectors to war, reaching far back into history. There have been the tax-resisters, who regard the payment of taxes to governments which commit the vilest of atrocities (not the least of which is the sponsorship of nuclear power) as a clear form of enslavement.

There are those who participated in the civil rights movement actively and courageously, in numbers which finally became too large and too effective to ignore totally. Encouraging too were the growing numbers who made some sacrifices to stop the American atrocity called the Vietnam War.

And most recently, there is the very encouraging sign of a vital and growing anti-nuclear movement, a movement which really seems to appreciate what the real issues are.

Nevertheless, the numbers who think justice is THE worthwhile human goal, who understand that power-systems and justice are never going to be compatible, are still small. That should not surprise us, when we remember that only about 100 years ago, overt slavery was practiced in this country; that 60 years ago, women were not even allowed to vote; that 20 years ago, the movies were still suggesting that *the American Indians* were the savages, rather than the whites who devastated them and their culture; and that until the last decade, this society treated its physically handicapped, mentally handicapped, and homosexual minorities like sub-humans.

What gives me hope about our enormous task is the *evidence* that insistence on human rights is growing among the American people.

And I think hostility toward those who violate human rights—for instance, by poisoning people—will grow very much more rapidly now that the corpses are appearing among the labor force, thanks to *chemical* poisons ruthlessly administered by job-givers, back when nuclear power was still just a gleam in Dr. Glenn Seaborg's eyes. Probably some of you saw the ABC-TV documentary on July 14 (1978) entitled "Asbestos: The Way to a Dusty Death", which revealed the collusive efforts to *cover up* and then to *ignore* the hazards of exposing workers to asbestos—with the result that perhaps more than a MILLION people are already condemned to a miserable, premature death. Such atrocities are not the exception; they are the norm of behavior by those with the power to use economic blackmail.

The Past Nine Years

About nine years ago, I realized that the story of asbestos was about to be repeated with radioactive poisons. There is not, nor will there ever be, a credible way to prevent the irreversible contamination of the Earth by radioactive poisons in a nuclear power economy. Nuclear power is simply incompatible with human health. That became obvious to me as a chemist and as a physician.

So in the early 1970's, I advocated that we should close down existing nuclear power plants and never build any more of them. Sometime later, Dr. Ralph Lapp—who has earned good consulting fees for teaching electric utility officials how to manage questions raised about nuclear power by the pesky public—honored me by stating that I was one nuclear critic who was "beyond the pale of reasonable communication". I have had compliments in my day, but never one so nice.

In the early 1970's, there were others who were concerned about nuclear power too. The focus of many of them, however, was on getting a specific nuclear plant constructed ELSEWHERE, just so that it is not near *me*. Efforts to get a murderous technology moved into the back-yard of *someone more ignorant*, I found nauseating.

However, I do like a proposal put forth by John Lane and Jay Kinney. If we were to insist that, if any of

Reprinted by courtesy of Jay Kinney*
and *In These Times*.

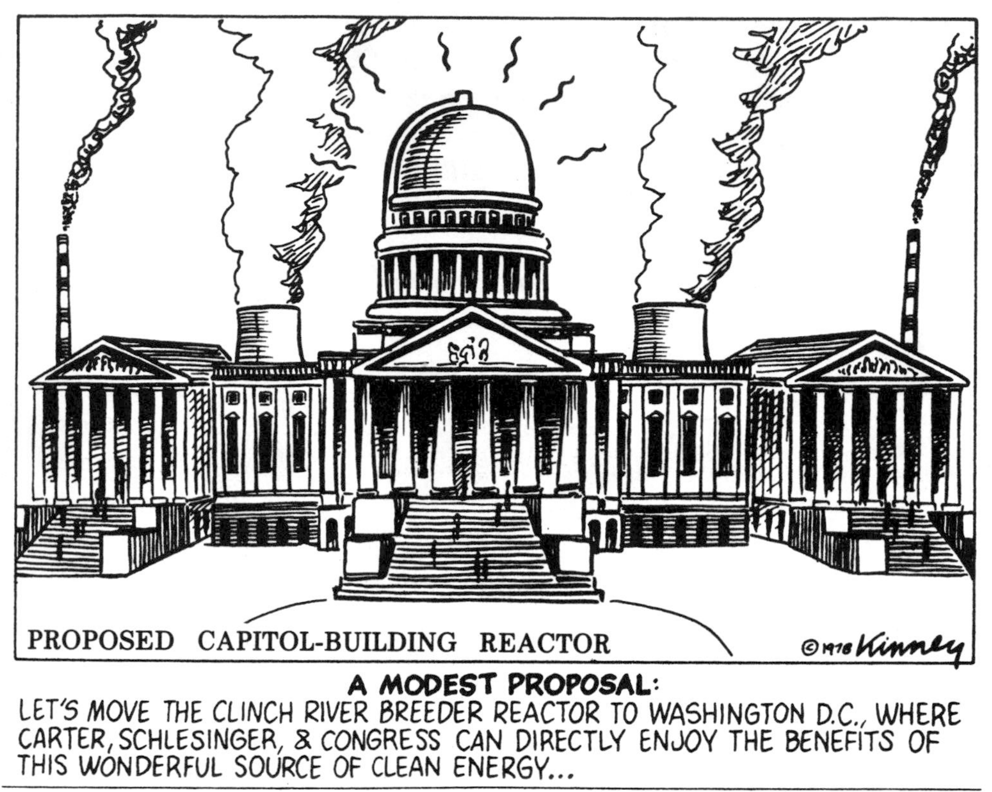

Jimmy Carter's proposed nukes are actually built, they must be built right in the middle of cities (which is where the power will allegedly be needed)—and especially sited right next to state capitol buildings where so many legislators waffle and bray about nuclear safety—we might rapidly end urban apathy on this issue, and expose the hypocrisy about nuclear plants as "good neighbors".

In the past nine years, there has been plenty of nonsense about being afraid to be labeled "anti-nuclear",

\+ + + + + + + + + + + + + +

* Jay Kinney is co-creator of the comic book *Cover-Up Lowdown*; Rip Off Press, POB 14158, San Francisco 94114.

with the resultant thrust of "We just want to make nuclear power SAFE", a message which is hard to distinguish from Mr. Megawatt's *own* claim. Nevertheless, that approach dominated the nuclear "safeguards" initiatives which went down to defeat two years ago.

In my opinion, one inadvertently fortifies the industry's *falsehood* that nuclear power CAN be safe and acceptable, by focusing on widget-fixing—getting stricter rules for radioactive transport, decent testing of the Emergency Core Cooling Systems, better evacuation plans in case of nuclear "incidents", improved security against nuclear terrorists (which means our assistance in creating a police-state), stepped up monitoring to track radioactive poisons AFTER they have been released irretrievably into our environment, bigger and better studies on waste burial as if we should contemplate *creating* more wastes to bury, and even getting different formats for letting citizens speak before nuclear licensing boards whose function is to *grant* more nuclear licenses! To me, upgrading the safety of existing and future nukes is good only if we submit to HAVING nukes, but I do not consent to having nukes at all.

Since it is the *totality* of the nuclear problems which tells us that nuclear power can never be acceptably safe, we surely *must* continue educating the public about them—but without falsely implying the problems are solvable. I realize that even the move-it-somewhere-else approach to nukes, in the early days of the movement, *did help* to start the process of public education, and the chain of events which led to the exposure of the lies concerning the cheapness and safety of nuclear power. And finally to an openly anti-nuclear movement.

The reason that I am heartened by the present anti-nuclear movement, of which the Trojan Decommissioning Alliance, the Abalone Alliance, the Crabshell, Palmetto, Clamshell, and several other alliances, are the cutting edge, is that you recognize nuclear power for what it is—a crime against humanity, with its built-in premeditated random murder of humans, even humans yet unborn, and a manifestation of injustice heaped upon humans by a social system worshiping at the shrine of power of *some* humans to do whatever they want to *other* humans. It is that association—nuclear power as a violation of justice—which is exciting and heartening. It is a big, big advance from widget-fixing.

From Widgets to Justice by Steps

To be sure, it is possible to educate new people only one step at a time, so it is fine that much effort has gone into teaching the public about the fantastic economic rip-off which nuclear power represents. And it is good that labor-people are being helped to see how nuclear power takes jobs away; it is useful to show that the reason we need to expand the economy, is because energy is *destroying* jobs so fast that an expanding economy is the only way to avoid massive unemployment! The mythology that we need to use more energy to make the economy healthy is being well demolished.

But *if* rip-offs were all that is at issue, the nuclear power controversy, and the exposure of the lies surrounding it, would not mean so much to me. It is the fact that a rapidly growing movement is being built of people who are concerned about justice *as a whole*, and the relationship of nuclear power to additional abominations, which is heartening.

I attended the Seabrook Rally on June 25 (1978). There were some 18,000 to 20,000 people there. Lest any of you be confused about the cancellation of a civil disobedience action in association with that event, let me assure you that Seabrook was *not* 18,000 or 20,000 people just there for a picnic. The thousands who sat in a hot, baking sun responding to talks hour after hour, clearly understood very well what the relationship is of nuclear power to human rights. You just could not have been at Seabrook without realizing that a truly meaningful anti-nuclear movement centering around justice is alive, flourishing, and growing at an astonishing pace.

Courtesy of Ray Collins of the *Seattle Post-Intelligencer.*

The meaningful anti-nuclear movement is on the cutting edge of a bigger movement to eliminate the disease of power-acquisition, and to build a world based instead upon justice and inalienable rights. Obviously the numbers have to grow immensely. We face two main obstacles to increasing our numbers:

Two Obstacles to Progress

(1.) ECONOMIC FEAR, which is based upon the threat that you and others will lose your jobs unless you go along with any and all abuses the power-system wishes to perpetrate upon you, from vinyl chloride poisoning, asbestos poisoning, herbicide poisoning, right through a long list to radiation poisoning.

(2.) FEAR OF ATTACK by another country, a fear which relates to the absurd belief that *countries*, rather than a few specific individuals, are responsible for wars; this widely held belief blocks the kind of clear thinking which might prevent nuclear holocaust.

Government is certainly *not* going to eliminate either obstacle to increasing our numbers. For one thing, the system does *not want* to get rid of unemployment. That should be quite obvious, since the system thrives on using the fear of unemployment. Since the government serves the system, and since unemployment is our system's essential tool for controlling people, it escapes me how some of my liberal friends can expect government *ever* to be constructive in eliminating the fear of unemployment.

Countering Economic Blackmail

I think it is up to us to explore some innovative ideas to counter economic blackmail. Many of you have made a great beginning by personally disconnecting from the larger economic system. But since that does not reduce the effect of economic blackmail on the great bulk of the population, it does not directly help increase the size of the justice movement. And quite obviously, "economic disconnect" does not provide a mechanism which can protect you, or your fellow humans, or your descendants, from victimization by nuclear war.

To counter economic blackmail, perhaps we need to give more thought to ways in which *we* could take care of people facing disemployment by the system, not by subsidizing their unemployment checks, but rather by *using* their capabilities. There must be desirable and desired services they can provide to people in the justice movement, in exchange for salaries or services provided to *them* by people in the justice movement.

If we assume that every reasonably healthy adult is able either to take care of his own needs, as can almost every other creature on Earth, or is able to exchange needed services with other humans, it is a deep mystery why we can't establish mini-economies based on affordable tools right in the middle of the bigger economy. The Amish community in Pennsylvania does do it successfully. After all, humans with hand-tools ARE CAPABLE of meeting their needs for food, shelter, clothing, and education—providing they are not prevented by sky-high property taxes from having the necessary land.

The paradox of capable humans who can not find work to do, is one which has apparently defied solution for many generations now. And not just in capitalist economies. Recently I read that Cuba has now developed an unemployment problem among its growing number of college graduates. Nevertheless, I think the unemployment paradox, and its concomitant economic blackmail, *can* be eliminated if they receive your clever attention.

With regard to employed people rustling up the money to employ people *who lose their jobs in the system*, I hope you will ponder upon the following figures. A report released last month by the White House's Office of Management and Budget states that Americans now spend about 700 million hours each year filling out federal forms, and that the estimated cost of preparing and processing all this paperwork is $100 billion a year. That's a reduction in useful income of about $500 for every man, woman and child in the nation, or $2,000 per family per year. But we don't "see" it because it is mostly hidden in the price of what we

buy. Government IS inflation, not an agency seeking to stop inflation. Just by eliminating *federal* forms, ten employed families could afford to hire one UNemployed family *to do useful work* for a salary of $20,000 per year!

Reducing the Power of Government

Since government actions have always ended up serving the power-elite, I consider government to be a big part of our problem and *not* the solution. And by taking more than one-third of our incomes away in various taxes, government drastically reduces our already meagre capabilities. Every tax-dollar paid to government is a guarantee of one dollar less available for trying to SOLVE problems. No doubt my liberal friends will disagree again, but disagreement has no power to alter the truth.

Can you imagine, even in your wildest flight of imagination, how paying the salary of Dr. James Schlesinger and his 10,000 employes is going to help *solve* anything? If you can believe that your taxes are helping to *solve* mankind's twin nuclear problems, then you can probably be blissfully narcotized into believing that SALT is a meaningful step toward preventing nuclear holocaust. I can not.

If government is not going to provide solutions, then it would seem logical that the sooner we have less government at all levels, the sooner we can free up some resources which *we* can direct toward solving problems. I would urge that you join, and help guide, the sentiment for tax-revolt in this country. There may be some unfamiliar bedfellows with you in that effort, but I would not assume that tax-revolt is either a racist backlash or a selfish backlash against the needy. For some, it surely is, but it is also an important backlash against tyranny. Virtual elimination of government would put enough money back in our hands so that we could be directly, voluntarily, and warmly generous toward the people who *do need help*. The sense of *personal* responsibility for our fellow humans would be encouraged instead of killed.

Coping with "National Security" Concerns

When we consider the second obstacle to increasing the numbers in the justice movement—namely, the widespread concern about our "national security"—we face a very, very difficult task. Because as long as FORCE is tolerated by peoples everywhere as a way for rulers to settle their conflicts, then both military secrecy and the striving to achieve a first strike are inevitable.

I think the American people as a whole understand at the gut level that self-defense *is a necessity* as long as rulers have military force at their command, and therefore I think the American people will continue tuning out when they hear unilateral disarmament steps advocated. I think their gut feeling about an external threat is *correct. Your* gut feeling, that the deterrent will not deter and that nuclear weapons mean genocide, is ALSO *correct.* Mankind would have a better chance if there were communication, instead of confrontation, between the so-called militarists and the so-called peaceniks.

In science, when two apparently incompatible observations are each real, true, and valid, we know an explanation *has to exist*. We can not pick the data that we *like* the best, and throw the rest away! Reconciling conflicting data can be exquisitely difficult, but the effort almost always leads to new insights. With respect to the nuclear arms race, I have been trying to reconcile the valid views of the militarists with the valid views of the peaceniks. I think the medical model of the power-disease *does* explain why both are valid. And I think the explanation is of more than academic interest, because the way one analyses the driving forces in any situation—scientific or political—makes a very big difference in what you decide to *do* in terms of action. I think we have to cope with the following logic:

> (1.) In order to eliminate *nuclear* weapons from this Earth, humans must also eliminate the use of force of ANY sort as an acceptable way to resolve conflicts. And,
>
> (2.) In order to eliminate the use of force, humans must keep the power-lovers OUT of positions of power.

Even if it were somehow possible to get power-wielders, worldwide, to renounce *nuclear* weapons, as long as power-lovers remain in charge of the world, they will—because of the nature of their disease—continue to harbor the idea of using FORCE in order to increase their power . . . and so the old ''R&D''(research and development) will just begin on things even worse than nuclear weapons. Like mind control. The problem is keeping the power-lovers OUT of power.

As I stated earlier, I believe interference with the nuclear capability of ONLY the United States would be an invitation to holocaust. I am certain *that* opinion rubs many of you the wrong way, but I urge you to think it through very carefully again. The possession of nuclear weapons certainly reflects the willingness to use them under the ''right'' circumstances, and we peons are not even offered a definition of ''right'' before genocide would be committed in our names. It is truly an abomination of the first rank in every way. Although the *possession* of nuclear weapons is certainly a violation of human rights, I think NOT possessing them at this time would result in an even greater violation of human rights, by inviting either nuclear incineration or Soviet enslavement of 3 billion additional people. Either we learn how to take military hardware and nuclear toys out of the hands of power elites EVERYWHERE, or we are getting nowhere on this problem.

Reaching Ordinary People Worldwide

Our toughest problem is that public enlightenment for this momentous change must be worldwide. It is essential to learn how to reach the Russian people, the Chinese people, and people everywhere. If our own government were truly committed to eliminating the use of force, we might find communication satellites and other effective technologies at the disposal of such an effort.

In this country, we still have the freedom to learn, and to speak freely to others. The Russians no longer have this opportunity, because the Soviet power-structure has adopted Hitler's concept of the Thousand Year Reich, replete with every available tool of body and mind control. But it is safe to assume that the Russian people share the same goals as humans anywhere else, and that they are victims of the Soviet power system, not its perpetrators.

John Sullivan. Chronicle Publishing Co., 1978.

Somehow we must create a way to get through to the Russian people and to help them join the effort to eliminate the disease of power-acquisition by some humans over others, just as we need to do here. How to be effective should occupy the best of your minds, for there is not too much time left.

It would be appropriate if today's nuclear weaponeers were in the forefront of this movement. They should be clamoring more loudly than anyone for steps to make their disgusting jobs unnecessary. But then, waiting for *their* leadership would be like waiting for the medical and legal professions to lead the fight against nuclear energy!

Quite in addition to the normal survival instinct which makes weaponeers act to *protect* their jobs, I can tell you something else which happens to weaponeers. They get carried away with the elegance of their scientific and technical insights, which are undeniably clever. *Often* I have heard them refer to a bomb-design as "neat," and even more often as "sweet." I have heard *that* even from J. Robert Oppenheimer, who opposed development of the hydrogen bomb . . . and who was persecuted heavily for his opposition.

Will the USA and USSR Merge?

Because nuclear holocaust is a near-certainty, for the reasons I have tried to present, it is possible that the rulers of the USA and USSR may come to regard this particular technology as too dangerous for *themselves*. If so, they could probably arrange for mutual nuclear disarmament quite easily. But I do not think they *would* get rid of nuclear weapons; it is far more likely that *their* solution to the problem would be to merge and to *consolidate* their nuclear forces in order to defend themselves from the rest of the world. As long as nuclear weapons can be made by the Third World nations, what chance is there of the USA and USSR giving

Robert Graysmith in *The San Francisco Chronicle*.

"Now we can destroy the world without destroying the world!"

them up? Merger of the super-powers might protect Americans from nuclear incineration, but merger would drag us into an even deeper pit. A society dominated by the terror of getting sent to the Soviet-style "gulag" prison system is truly the antithesis of justice and freedom, and the *consolidation* of power instead of the *elimination* of power would virtually guarantee that there could never be a triumph for the justice-seekers.

Since merger of the super-powers would be such a nauseating way to eliminate our risk of nuclear incineration, we must devise other approaches.

The Nuclear Problems on a Scale of Injustice

I am trying to say it as I see it. The way I see reality is not encouraging. It is difficult indeed to know what to do. But the fact that a problem is difficult, just does not make it attractive to go solve some non-problems instead! In the face of enormous problems, I feel that time spent thinking is *well* spent. Activity for activity's sake can turn out to be wheel-spinning or widget-fixing.

One can always deceive oneself that small steps leading nowhere will somehow arrive at the trickiest, most difficult destination imaginable, but let us *not* retreat into such self-deception. Let us try to distinguish between the small random steps which lead nowhere, and the small steps which *do* advance toward an important destination. If the human species is worth worrying about, and if it has a lot of enormous problems which share a common cause, let us always ask ourselves if our steps are zeroing in on that cause.

In closing, I must ask a question. Why do we consider nuclear weapons to be the worst kind of monstrosity? Yes, 70 million Americans could get wiped out practically overnight, plus some fallout victims and descendants later. That is obviously monstrous. But compared to WHAT? Robert McNamara, President of the World Bank, has tried to tell us (October, 1976) that there are 900 MILLION "severely deprived human beings struggling to survive in a set of squalid and degrading circumstances almost beyond the power of our sophisticated imaginations and privileged circumstances to conceive . . . . Malnutrition saps their energy, stunts their bodies, and shortens their lives. Illiteracy darkens their minds and forecloses their futures. Squalor and ugliness pollute and poison their surroundings."

Courtesy of Jack M. McLeod and *The Buffalo Evening News.*

Underdeveloped Nations' Arms Race

Including all ages, 70 million people worldwide face starvation *every year*, in an annual holocaust few Americans even know about, and 400 to 600 million people on Earth have diets inadequate to sustain normal brain development. That is an even MORE monstrous injustice than nuclear war. Isn't it?

Getting killed is something we regard as an injustice, and we are mad about it, whether the killing is done to us *fast* with bombs, or *slowly* with nuclear pollution from civilian nuclear energy. But if we were to fight nuclear power and nuclear weapons only out of concern for our *own*

Graphic by Meg Simonds, People against Nuclear Power and The Abalone Alliance.

survival, while ignoring the murderous abuse of hundreds of millions of our fellow humans in the Third World, we would essentially be saying that it is more important to save life for Americans than it is to save life for Latin Americans, Africans, or Asians. It is *not* more important. To worry about the survival and health of just unborn *American* generations would be only a small advance beyond the so-called concern of those anti-nuclear activists who faded from the fight for justice as soon as the nuke planned for *their* backyard was canceled!

What I find so encouraging about the recent anti-nuclear movement is that its activists—all of you—seem to see the anti-nuclear movement in its larger context, on the cutting edge of a far, far bigger movement for justice throughout the whole human species and toward the other species on this planet. It just happens that we are learning how to make our contribution to this movement via the nuclear issues.

It is in that context that I regard what you are doing as having *immense* importance for all humanity. NO NUKES!

□ □ □ □ □

9.

LAW VERSUS JUSTICE:

What's Happening at the Trials of the "Fence-Jumpers"?

It is my purpose here to examine what I consider some of the most profound threats to freedom and justice which our society has yet witnessed. In order to clarify how such a serious claim comes to be discussed in relation to ways of opposing nuclear power, I must go over a bit of recent history.

The Tower-Toppling Trial

In 1974, a young Amherst graduate and communal farmer, Sam Lovejoy, chose to express his opposition to nuclear power as an abominable threat to humanity by toppling a meteorological tower in Massachusetts, said tower having been constructed to obtain data needed for later construction of a nuclear power plant. He chose to admit his act of "civil disobedience" on Washington's Birthday. Having toppled the tower, he promptly turned himself into the authorities, finally convinced them that he had indeed toppled the tower, and managed to get the authorities to indict him for this apparently dastardly act.

In due course, he came to trial, where he served as his own attorney, with the obvious purpose of putting nuclear power on trial. In order to carry through his purpose, he needed "expert" testimony concerning the hazard of nuclear power to health. He asked me to testify on his behalf, and I agreed that I would.

When it came time for Sam to call me to the witness stand, he stated that he wished to have the hazards of nuclear power explained, and promptly the prosecuting attorney objected that the hazards of nuclear power were irrelevant, since the issue was whether or not he had maliciously destroyed property, not whether nuclear power was dangerous.

The Judge seemed a bit perplexed about ruling on the prosecutor's objection. The course he took was to let Sam conduct a direct examination of me for about 1½ hours, *but without the jury being present.* Then he adjourned the Court for a period while he considered whether the jury should or should not have the opportunity to hear my testimony concerning the health hazards of nuclear power. Finally he returned to the courtroom and announced that he would *not* let the jury hear my testimony; in other words, he sustained the prosecutor's objection.

But the Judge did something extraordinary in addition to denying the jury the opportunity to hear about the hazards of nuclear power. He stated to the Court that in the event that Sam Lovejoy was found guilty, he would not pass sentence but would instead refer the case to the Massachusetts Supreme Court to decide whether he had erred in denying the jury the opportunity to hear and consider the evidence concerning nuclear power in reaching its decision on Sam Lovejoy's guilt or innocence. As it turned out, that issue never came up because the entire charge against Sam was thrown out on a faulty indictment basis (1).

There was something about that Judge which I thought then was profound. In the four years which have passed, I have given the matter a great deal of thought, and I am now certain that my initial impression that this Judge was a serious thinker, was indeed correct. Clearly something was troubling the Judge profoundly in reaching his decision to withhold the evidence concerning nuclear power from the jury. A lesser jurist could simply have sustained the prosecutor's objection. A lesser jurist could simply never have made that statement concerning sentencing and the possibility that he had erred. Surely judges are not popular with the Establishment if they do *not* hold property rights as paramount and far more important than human life or health.

I'm glad that our society considers private property-rights to be sacred. But also I hold that the right of

\+ + + + + + + + + + + + + +

(1) "Lovejoy's Nuclear War", a one-hour documentary film about the case and about local public opinion toward Sam, has won several national film-awards; details from Green Mountain Post Films, Box 177, Montague, MA 01351.

private property (like the right of free speech) is *forfeited* by people who use it to deny another sacred right, namely someone's inalienable right to *life*. It is ludicrous to suppose that an electric utility which is using (or proposes to use) its property to commit premeditated random *murder*, retains all the sacred rights of private property! Likewise, it is ludicrous to suppose that the Nazi Party, which has not only advocated but also committed genocide, retains the right of free speech. It would make a mockery out of the right to life, to allow either the right of free speech or the right of private property to be used *to deny the right of life* to anyone.

I believe it is interesting and important that the Lovejoy trial was held in New England. I say this because I rather suspect that New England is one of those rare places in the United States where people still have a recollection that there ever existed something called the "Declaration of Independence".

Something was indeed bothering that Judge, something which prevented him from a *summary* dismissal of Sam's right to have nuclear power discussed before the jury in its evaluation of his behavior in toppling the tower. The more that I have thought about this episode in the past four years, the more convinced I have become that this Judge recognized that *justice and the law are not synonymous*, and that justice was really of importance. Not many people ever think about this distinction, or even admit that justice and the law are two separate entities.

The Issue: Human Rights

I do not now hold, nor have I held in the past, any opinion one way or the other concerning whether "civil disobedience" is a good or bad approach to fighting nuclear power. I simply do not know. There is no doubt in my mind that the promotion of nuclear power is itself a criminal act of the worst sort, given the Constitution under which we live. No doubt at all.

The Constitution of the United States does not permit the taking of life without due process of law. Nuclear power, which begins its random murder of citizens of the United States even before the nuclear plant goes into operation, is obviously an infringement of Constitutional rights. Even if there were no Constitutional violation, nuclear power is a violation of natural rights and justice.

There are those in our society who don't realize that justice is a concept which exists quite independently of any laws made by legislatures—even though that independence is commonly acknowledged whenever someone says, "That law is unfair".

Moreover, probably most Americans have forgotten the Ninth Amendment to our Constitution, stating the fact that certain rights inhere naturally to the people whether or not the Constitution spells them out specifically. *The right to life* is certainly one such natural right.

I mentioned above that nuclear power starts to commit murder even before the plant goes into operation. It does so by guaranteeing that people are going to be poisoned for hundreds of thousands of years by radon and its daughter products brought to the surface of the earth in the course of mining the uranium needed to operate the nuclear plants. Had these substances remained in the bowels of the Earth, they would have done no harm.

And the random murder of citizens is further increased by the fact that there are emissions of radioactive substances both in the normal and abnormal operation of the nuclear fuel cycle. Just how large the number of people to be murdered is, depends upon the outcome of the largest experiment upon humans that evil genius has yet concocted. I say that this experiment is evil in the extreme because it not only kills humans of this generation, but also because it reaches into countless future generations with its lethal effects. That certainly qualifies as the depth of depravity, particularly because future generations may have much better common sense than ours does, and might not elect to

By Janet Kailin, Sequim WA 98382.

be poisoned if they had the opportunity to decide for themselves.

For our purposes here, it is immaterial whether the numbers to be murdered by the operation of a single plant, such as the Palo Verde plant, are to be counted in the tens, the hundreds, or the millions. Murder is murder! We have not yet adopted the position that criminals who kill people have to kill *a certain number* before their action qualifies as a crime.

It may indeed be true that society could reach the decision that it is all right to murder a certain number of people in the course of industrial activities, but that is a profound decision and should only be taken with full cognizance of its true meaning.

Law vs. Justice

It is said that nuclear power plants can operate *legally* simply because they are licensed to operate by the Nuclear Regulatory Commission. The Nuclear Regulatory Commission is operating *legally* because Congress legislated it into existence to issue such licenses. But what has all this to do with justice and natural rights? Congress has no authority under the Constitution to issue murder licenses. Moreover, Congress could have no such authority, simply because one of the rights protected by the Ninth Amendment is the *natural right to justice and to life*.

That is my opinion, and it would not be altered one whit if there were 100 decisions by the Supreme Court which stated that it is permissible to murder people. There is a higher law.

It amazes me that people don't seem to realize the implications of permitting laws to be passed which violate justice and natural rights. It amazes me especially since it is so soon after the Nazi Holocaust and the Nuremberg Trials. In Nazi Germany the rulers, as evil people as one can imagine, wished to carry out a program of genocide. Because of the recognition that people *might* object to such a gross violation of justice and human rights, even the Nazis decided to make the process legal, at least in part, by passing laws which permitted judges to send people to their death with no justification at all other than a Nazi-passed law.

At the Nuremberg Trials, the United States declared that this sham of using *"laws"* to subvert *justice* was a heinous crime, and we meted out severe sentences to judges who had used the Nazi "laws" as a shield for the crimes which they (the judges) committed on the bench.

If the Congress of the United States can permit the Nuclear Regulatory Commission to deprive people of life without due process of law, and if the Supreme Court turns its head from realizing this, as it did in declaring the Price-Anderson Act to be Constitutional, where are the guarantees that far worse injustices and violations of human rights will not be carried through in the future?

Personal Responsibility

In the USA, we have already accepted the policy of experimentation on involuntary human subjects. Every year, we introduce new chemical compounds of *uncertain* toxicity into the workplace and the biosphere. In the mid-fifties—when the toxity of low-dose *radiation* was still uncertain—we were testing nuclear bombs in the atmosphere and launching the Atoms for Peace program.

It should have been clear to me, even then, that both atmospheric bomb-testing and nuclear power constituted experimentation on involuntary human subjects, indeed on all forms of life. Instead, I am on record in 1957 as *not* being worried yet about fallout, and still being optimistic about the benefits of nuclear power.

There is no way I can justify my failure to help sound an alarm over these activities many years sooner than I did.

I feel that at least several hundred scientists

trained in the biomedical aspect of atomic energy—myself definitely included—are candidates for Nuremberg-type trials for crimes against humanity through our gross negligence and irresponsibility.

The Fence-Jumpers . . .

Now that we *know* the hazard of low-dose radiation, the crime is not experimentation—it's *murder*. Perhaps the "fence-jumpers" at nuclear power sites are the Paul Reveres of today, as they use their bodies to try warning us against *accepting* a policy of premeditated random murder.

"Must the citizen ever for a moment, or in the least degree, resign his conscience to the legislator? Why has every man a conscience, then? I think that we should be men first, and subjects afterwards.

"It is not desirable to cultivate a respect for the law, so much as for the right. The only obligation which I have a right to assume is to do at any time what I think right . . . Law never made men a whit more just; and, by means of respect for it, even the well disposed are daily made the agents of injustice . . .

"The mass of men serve the state thus, not as men mainly, but as machines, with their bodies . . . In most cases there is no free exercise whatever of the judgment or of the moral sense; but they put themselves on a level with wood and earth and stones; and wooden men can perhaps be manufactured that will serve the purpose as well. Such command no more respect than men of straw or a lump of dirt . . . Yet such as these even are commonly esteemed good citizens.

"Others—as most legislators, politicians, lawyers, ministers, and office-holders—serve the state chiefly with their heads; and, as they rarely make any moral distinctions, they are as likely to serve the Devil, without *intending* it, as God."

Henry David Thoreau's
"Civil Disobedience", 1849, in
the magazine, THE DIAL.

But apparently public apathy has something to do with *numbers*. If the numbers who die from nuclear power are "not too large", particularly if they die quietly without sporting a little flag saying "I am a victim of nuclear power", there is apparently nothing to alert people to their imminent danger of losing freedom, justice, and the inalienable right to life.

What would people think about Congress setting up a new Commission to permit licenses to be issued for a totally novel industry which would kill 1,000 persons per year? Would the people be concerned? What about 10,000 persons per year? What about 100,000 persons per year?

It is obvious to any rational person that when you permit the first murder-license, you have opened the floodgates. I say this fully cognizant that, one day, society may choose to permit the murder of some number of people to achieve a so-called "benefit". But one had better think *long and hard* about how such permission becomes permitted, or the consequences could be the loss of all freedom and justice in our society.

Where and when should people begin to raise questions about the subversion of justice and natural rights by laws passed by fallible men, and declared "Constitutional" by fallible justices on the bench? It is my opinion that the Judge in the Sam Lovejoy case was having trouble with this question.

. . . And Their Trials

Recently, as just about everyone knows, there have been a number of "civil disobedience" actions against nuclear power facilities. Fence-jumping. Occupations. Blockades. In all these cases, it is alleged that "criminal trespass" upon private property has occurred, and the "disobedients" are being brought to trial for their alleged crimes.

In almost all cases thus far, the judges have held that expert testimony concerning the hazard of nuclear power *cannot be presented to the jury*. Stated succinctly, the action of the judges means that one is not entitled to present to a jury of his peers the reasons for his action.

But a person accused of a crime *must* be allowed to explain, with the help of expert witnesses when necessary, in what way his behavior was true to a law *higher* than the statute. When a judge denies the jury's right to hear and then decide for *itself* whether a "higher law" was really applicable to the alleged crime, the judge is denying the right (and duty) of jury-members and all other members of society to consider the higher laws—by which I mean the very principles of justice and human rights from which laws are meant to be derived.

When juries lose the right to decide where justice lies, when juries must confine themselves to statutes, then our legal system is morally bankrupt and restraints on tyranny are gone. In earlier times, juries could be defenses against the king's tyranny because—while the king could still make any unjust laws he wanted—juries did not have to follow them. Recent rulings which deny the relevance of a law higher than the statutes, set us back to the Nazi era, or even to medieval times—except that the king has been replaced by a few hundred legislators whose "principles" are rated below those of most snake-oil salesmen by many Americans.

At Nuremberg, we said that individuals have "duties which transcend the national obligations of obedience imposed by the individual state". Now our own judges are denying it, by preventing fence-jumpers from showing juries how a higher law was involved in their behavior.

I am totally indignant when some people

suggest that we have to have a major *war* before we can invoke the Nuremberg Principles! Those principles are simply a statement of the obvious: there is a higher law, a standard of justice and morality, which enables us to evaluate particular statutes and to guide our own behavior, and consequently the higher law has a greater claim on our allegiance than the lower laws. To assert that we have to have massive carnage and "victory" before the Nuremberg Principles apply, is an absurdity. The Nuremberg Principles are statements for all men and all times.

In some states, there is a law which holds that a defense is valid if it claims that a violation of law was committed to prevent a *greater* harm than the harm caused by the law-breaking. Other states do not have this statute. But I do not think the existence or non-existence of the "competing harms" statute is of any relevance whatever. It is simply obscene that a judge in any state, with or without such a statute, can prevent a person from telling a jury of his peers why he committed an act for which he is being tried. That is simply unjust on the face of it, no matter what the legal doctrine holds.

Law Profession: Unaware? or Something Worse?

At first I thought it might be just my ignorance of jurisprudence which made me fail to understand why it must be reasonable and just for judges to prevent the juries from hearing how nuclear power violates human rights. It did not seem just to me, but I thought perhaps I simply did not understand.

Then I thought back on that Judge in the Lovejoy case. He realized that the issue was of some consequence. Moreover, I have talked with the lawyers for many of the "civil disobedient" cases, and uniformly they tell me *they* are incensed by such action of judges. So if lawyers find fault with the action of the judges, if the Judge in the Lovejoy case was concerned lest he had erred in preventing the jury from hearing the evidence about nuclear power, then there must indeed be a real issue here even to those who know the ins and outs of the "law".

The issue seems to me so profound that I would think the entire legal profession would be up in arms about it—that is, if the legal profession is concerned about *justice*. Perhaps it has not yet come to the profession's attention, in which case citizens should be asking every lawyer they know what he plans to do about this issue.

If it has come to lawyers' attention, and they are still doing nothing about this issue, one had better ask himself how much faith he wishes to place in the legal profession to protect freedom and justice. Maybe that is no concern of the legal profession.

Protecting *Our* Dissidents

The Palo Verde Nuclear Plant is going to commit random murder of citizens in Arizona—indeed its murders won't, in the long run, be limited to Arizona. That is sad. But it is less important, by far, in my opinion, than the implications of the current scene for the murder of natural rights and justice in the United States.

The experience of Nazi Germany has taught us that rulers can use legislative bodies to pass laws which will "justify" the crime of genocide, the greatest subversion of justice and human rights. Even a cursory examination of what is going on in the Soviet Union in such cases as that of Anatole Shcharansky shows that law is being used to violate justice there in a similar fashion.

Americans should indeed treasure the freedoms we have, for they are so rare in the world. What is so dangerous is *denial* of American rights (described in this paper) to some of our dissidents and to some of our juries.

While a path of great length separates the denial of rights now in the U.S. from denial of rights in Nazi Germany and Soviet Russia, the path is a direct one—a continuum—and its distance will be more and more easily traveled if we fail to stop transgressions against American rights at their earliest stages.

The reason is simple: by the time transgressions have become flagrant enough to disgust or even to threaten all decent people, the price of resistance has gone up exponentially.

The story is NOT this: It's this:

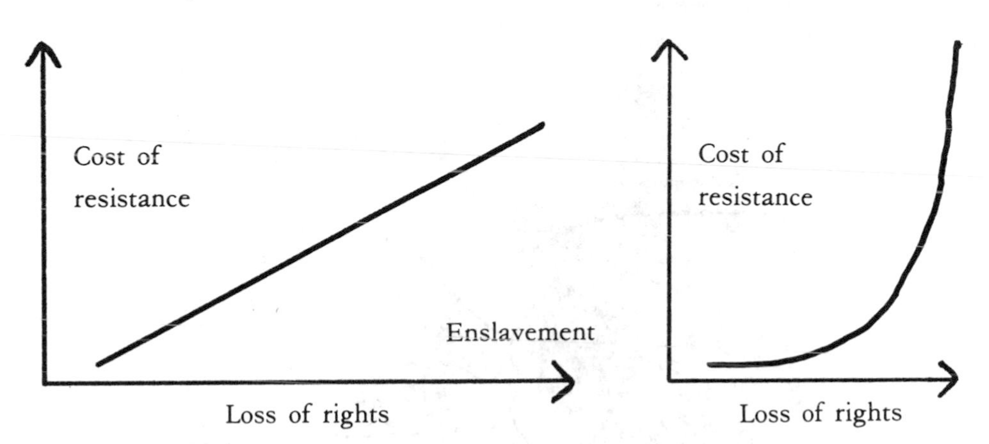

As the cost of resistance escalates from mere ridicule (as an "alarmist" and "exaggerator") to loss of jobs, promotions, and grants, resistance dwindles to insignificance. By the time resistance would have cost decent Germans their lives, virtually no resistance was made.

Transgressions against the human rights of *our* dissidents and against the rights of *our* juries have a significance, *whether or not* we choose to recognize it.

Freedom from slavery (such as that of Nazi Germany or the Soviet Union) and the preservation of human rights and justice, are issues of an importance which transcends by far the murders to be caused by nuclear power. It has been said, properly, that eternal vigilance is the price of freedom. I believe the current handling of the civil disobedience cases in the courts should be a warning that citizens had better have a hard look at how far we are going down the path to the loss of freedom and all our natural (human) rights by a blind obedience of the courts to "laws" which violate freedom, human rights, and justice.

If citizens do not insist on the higher law which takes precedence over all man-made laws, there is really not much left. We understood that at Nuremberg. We told the world that we understood that. How have we forgotten so soon?

By S. Gross. Reprinted by permission of the Chicago Tribune-New York News Syndicate.

□ □ □ □ □

From

AN ESSAY ON

• • • • • • • • • • • • • •

THE TRIAL BY JURY

• • • • • • • • • • • • • • • • • • • •

by Lysander Spooner

Few, if any, of the ideas in this book are brand new. In spite of non-acceptance, the same ideas have kept recurring to various humans throughout history. I find it impossible, with respect to the trials of the "fence-jumpers", to improve upon the essay (first published in 1852) by the lawyer, Lysander Spooner; so I shall quote below from its beginning and from its end. * *JWG.*

For more than six hundred years — that is, since Magna Carta, in 1215 — there has been no clearer principle of English or American constitutional law, than that, in criminal cases, it is not only the right and duty of juries to judge what are the facts, what is the law, and what was the moral intent of the accused; *but that it is also their right, and their primary and paramount duty, to judge of the justice of the law, and to hold all laws invalid, that are, in their opinion, unjust or oppressive, and all persons guiltless in violating, or resisting the execution of such laws.*

Unless such be the right and duty of jurors, it is plain that, instead of juries being a "palladium of liberty" — a barrier

\+ + + + + + + + + + + + + +

* The complete 220-page essay, and other essays by Spooner, are available from Laissez Faire Books, 206 Mercer St., New York City 10012. *Jury* is $6.00 including postage. Laissez Faire Books is a bookstore which specializes in books on liberty.

against tyranny and oppression of the government — they are really mere tools in its hands, for carrying into execution any injustice and oppression it may desire to have executed.

But for their right to judge of the law, *and the justice of the law*, juries would be no protection to an accused person, *even as to matters of fact*; for, if the government can dictate to a jury any law whatever, in a criminal case, it can certainly dictate to them the laws of evidence. That is, it can dictate what evidence is admissible, and what inadmissible, *and also what force or weight is to be given to the evidence admitted*. And if the government can thus dictate to a jury the laws of evidence, it cannot only make it necessary for them to convict on a partial exhibition of the evidence rightfully pertaining to the case, but it can even require them to convict on any evidence whatever that it pleases to offer them.

[Trial by jury] was anciently called "trial *per pais*" — that is, "trial by the country" . . .

It is supposed that, if twelve men be taken, *by lot*, from the mass of the people, without the possibility of any previous knowledge, choice, or selection of them, on the part of the government, the jury will be a fair epitome of "the country" at large, and not merely of the party or faction that sustain the measures of the government; that substantially all classes of opinions, prevailing among the people, will be represented in the jury; and especially that the opponents of the government (if the government have any opponents), will be represented there, as well as its friends . . .

It is fairly presumable that such a tribunal will agree to no conviction except as *substantially the whole country* would agree to, if they were present, taking part in the trial. A trial by such a tribunal is, therefore, in effect, "a trial by the country." In its results it probably comes as near to a trial by the whole country, as any trial that it is practicable to have, without too great inconvenience and expense.

And as unanimity is required for a conviction, it follows that no one can be convicted, except for the violation of such laws as substantially the *whole* country wish to have maintained. The government can enforce none of its laws . . . except such as substantially the whole people wish to have enforced. The government, therefore, consistently with the trial by jury, can exercise no powers over the people (or, what is the same thing, over the accused person, who represents the rights of the people) except as substantially the whole people of the country consent that it may exercise. In such a trial, therefore, "the country", or the people, judge of and determine their own liberties against the government, instead of the government's judging of and determining its own powers over the people.

But all this "trial by the country" would be no trial at all "by the country", but only a trial by the government, if the government could either declare who may, and who may not, be jurors, or could dictate to the jury anything whatever, either of law or evidence, that is of the essence of the trial . . .

. . . if the government may dictate to the jury *what laws they are to enforce*, it is no longer a "trial by the country", but a trial by the government; because the jury then try the accused, not by any standard of their own — not by their own judgments of their rightful liberties — but by a standard dictated to them by the government . . . In short, if the jury have no right to judge the justice of the law of the government, they plainly can do nothing to protect the people against the oppressions of the government; for there are no oppressions which the government may not authorize by law . . .

The question, then, between trial by jury, as thus described, and trial by government, is simply a question between liberty and despotism. The authority to judge what are the powers of the government, and what the liberties of the people, must necessarily be vested in one or the other of the parties themselves . . . If the authority be vested in the government, the government is absolute, and the people have no liberties except such as the government sees fit to indulge them with. If, on the other hand, that authority be vested in the people, then the people have all liberties (as against the government), except such as substantially the whole people (through a jury) choose to disclaim; and the government can exercise no power except such as substantially the whole people (through a jury) consent that it may exercise.

The force and justice of the preceding argument cannot be evaded by saying that the government is chosen by the people . . . and that to allow a jury, representing the people, to invalidate the acts of government, would therefore be arraying the people against themselves.

. . . in a representative government, there is no

absurdity or contradiction, nor any arraying of the people against themselves, in requiring that the statutes or enactments of the government shall pass the ordeal of any number of separate tribunals, before it shall be determined that they are to have the force of laws.

Our American constitutions have provided five of these separate tribunals, to wit, representatives, senate, executive, jury, and judges; and have made it necessary that each enactment shall pass the ordeal of these separate tribunals, before its authority can be established by the punishment of those who transgress it. And there is no more absurdity or inconsistency in making a jury one of these several tribunals, than there is in making the representatives, or the senate, or the executive, or the judges, one of them. There is no more absurdity in giving a jury a veto upon the laws, than there is in giving a veto to each of these other tribunals . . .

Neither is it of any avail to say, that, if the government abuse its power, and enact unjust and oppressive laws, the government may be changed by . . . the exercise of the right of suffrage . . . It can be exercised only periodically . . . Besides, when the suffrage is exercised, it gives no guaranty for the repeal of existing laws that are oppressive, and no security against the enactment of new ones that are equally so. The second body of legislators are liable and likely to be just as tyrannical as the first. If it be said that the second body may be chosen for their integrity, the answer is, that the first were chosen for that very reason, and yet proved tyrants. The second will be exposed to the same temptations as the first, and will be just as likely to prove tyrannical. Who ever heard that succeeding legislatures were, on the whole, more honest than those that preceded them? . . .

. . . If it be said that the first body were chosen for motives of injustice, that fact proves that there is a portion of society who desire to establish injustice; and if they were powerful enough or artful enough to procure the election of their instruments to compose the first legislature, they will be likely to be powerful or artful enough to procure the election of the same or similar instruments to compose the second . . .

The trial by jury, then, gives to any and every individual the liberty, at any time, to disregard or resist any law whatever of the government, if he be willing to submit to the decision of a jury, the questions: whether the law be intrinsically just and obligatory? and whether his conduct, in disregarding or resisting it, were right in itself? And any law, which does not, in such trial, obtain the unanimous sanction of twelve men, taken at random from the people, and judging according to the standard of justice in their own minds, free from all dictation and authority of the government, may be transgressed and resisted with impunity, by whomsoever pleases to transgress or resist it.

The trial by jury authorizes all this, or it is a sham and a hoax, utterly worthless for protecting the people against oppression. If it do not authorize an individual to resist the first and least act of tyranny, on the part of the government, it does not authorize him to resist the last and the greatest. If it do not authorize individuals to nip tyranny in the bud, it does not authorize them to cut it down when its branches are filled with the ripe fruits of plunder and oppression . . .

The principal objection, that will be made to the doctrine of this essay, is, that under it, a jury would paralyze the power of the majority, and veto all legislation that was not in accordance with the will of the whole, or nearly the whole, people.

The answer to this objection, is, that the limitation which would be thus imposed upon the legislative power . . . is the crowning merit of the trial by jury . . .

There is no particle of truth in the notion that the majority have a *right* to rule, or to exercise arbitrary power over, the minority, simply because the former are more numerous than the latter . . . And no more tyrannical principle was ever avowed, than that the will of the majority ought to have the force of law, without regard to its justice; or, what is the same thing, that the will of the majority ought always to be presumed to be in accordance with justice. Such a doctrine is only another form of the doctrine that might makes right . . .

[With trial by jury,] all legislation would be nullified, except the legislation of that general nature which impartially protected the rights, and subserved the interests, of all. The only legislation that could be sustained, would probably be such as tended directly to the maintenance of justice and liberty . . .

In short, government in practice would be brought to the necessity of a strict adherence to natural law, and natural justice, instead of being, as it is now, a great battle, in which avarice and ambition are constantly fighting for advantages and obtaining advantages over the natural rights of mankind.

□ □ □ □ □

"Your Honour, seven of us find the defendant guilty as charged. Three of us find him as guilty as they come, and two of us find him guilty from the word 'go'."

We can't win them all, at first. However, the situation is far from hopeless. When twenty anti-nuclear activists ("radio-activists") were tried for criminal trespass at the Zion nuclear power plant near Chicago, Judge Alphonse Witt *did* permit the defendants to explain why they thought a higher law applied, and to present the testimony of an expert witness to the jury. On January 29, 1979, the jury *acquitted* all the defendants.

JWG.

Another Way to Demonstrate Your Concern

You can put a copy of this book into your public or college library, or into the hands of a friend, just by sending the price of the book to the Committee, with the address of the intended recipient. The Committee will pay for the packing and postage. Please mail your coupon(s), or a list, with $3.95 per book, to:

Committee for Nuclear Responsibility
Main P.O.B. 11207
San Francisco, CA 94101

To:

With the compliments of:

To:

With the compliments of:

cutting lines

Citizen-groups and individuals may obtain *full boxes* of books (16 books per box) at a substantial discount. Inquiries are welcome, and will be promptly answered.

Courtesy of Ray Collins
of the *Seattle Post-Intelligencer*.

Location of Some Technical Reports

THE GOFMAN-TAMPLIN REPORTS

In late 1969 and 1970, Drs. Gofman and Tamplin issued a series of technical reports (known as "The G-T Series") documenting in detail the much larger association of ionizing radiation with cancer-causation than had been previously estimated. These reports examine all the major sources of data from known, human exposures (including certain medical therapies, the Hiroshima-Nagasaki data, the uranium miners, the radium dial-painters) as well as relevant data from animal exposures.

These reports, which are no longer available from the Committee, were submitted as testimony to two Congressional Committees, and are available at large libraries which keep collections of Congressional Hearings. Three volumes are involved, and the citations given below are correct. It is no error that testimony to one Committee is published in the Hearings of a competing Committee; nor is it an error that the date given to an entire volume may be earlier than the dates on materials included in that volume. Congress can operate that way.

ENVIRONMENTAL EFFECTS OF PRODUCING ELECTRIC POWER, Hearings before the Joint Committee on Atomic Energy, 91st Congress, first session, Part I, October and November 1969:

- "Low Dose Radiation, Chromosomes, and Cancer", presented at the 1969 IEEE Nuclear Science Symposium, San Francisco, October 29, 1969; pages 640-52.
- "Federal Radiation Council Guidelines for Radiation Exposure of the Population at Large — Protection or Disaster?", testimony presented to the Senate Committee on Public Works, November 18, 1969; pages 655-683.
- "Studies of Radium-Exposed Humans: The Fallacy Underlying a Major Foundation of NCRP, ICRP, and AEC Guidelines for Radiation Exposure to the Population-at-Large", supplement to testimony November 18, 1969; pages 695-706.

UNDERGROUND USES OF NUCLEAR ENERGY, Hearings before the Subcommittee on Air and Water Pollution of the Committee on Public Works, U.S. Senate, 91st Congress, first session, on S.3042, Part I, November 18-20, 1969:

- "A Proposal for at Least a Ten-Fold Reduction in the Federal

Radiation Council Guidelines for Radiation Exposure to the Population-at-Large; Supportive Evidence'', presented to the Joint Committee on Atomic Energy, January 28, 1970; pages 319-325.

- ''Studies of Radium-Exposed Humans II: Further Refutation of the R.D. Evans' Claim that the Linear, Non-Threshold Model of Human Radiation Carcinogenesis Is Incorrect'', supplement to testimony January 28, 1970; pages 326-350.
- ''The Colorado Plateau: Joachimsthal Revisited? An Analysis of the Lung Cancer Problem in Uranium and Hardrock Miners'', supplement to testimony January 28, 1970; pages 351-377.
- ''Radiation-Induction of Human Breast Cancer'', supplement to testimony January 28, 1970; pages 378-388.
- ''Radiation-Induction of Human Lung Cancer'', supplement to testimony January 28, 1970; pages 389-399.
- ''The Mechanism of Radiation Carcinogenesis'', supplement to testimony January 28, 1970; pages 400-418.
- ''ICRP Publication 14 vs. the Gofman-Tamplin Report'', supplement to testimony January 28, 1970; pages 419-425. (ICRP stands for International Commission on Radiological Protection).
- ''Major Fallacies in the AEC Staff Comments on the Gofman-Tamplin Papers and Congressional Testimony'', supplement to testimony January 28, 1970; pages 426-433.
- ''Radiation-Induction of Breast Cancer in the Rat; A Validation of the Linear Hypothesis of Radiation Carcinogenesis over the Range 0-600 Rads'', supplement to testimony January 28, 1970; pages 434-441.
- ''Radiation Aging by High LET Radiation: The Implications of Assuming Cell Nucleus Irradiation Is the Relevant Parameter'', by Donald P. Geesaman, supplement to testimony January 28, 1970; pages 442-444.

UNDERGROUND USES OF NUCLEAR ENERGY, Hearings before the Subcommittee on Air and Water Pollution of the Subcommittee on Public Works, U.S. Senate, 91st Congress, second session, on S.3042, Part II, August 5, 1970:

- ''16,000 Cancer Deaths from FRC Guideline Radiation (Gofman-Tamplin) vs. 160 Cancer Deaths from FRC Guideline Radiation (Dr. John Storer): A Refutation of the Storer Analysis'', testimony to the Joint Committee on Atomic Energy, February 9, 1970; pages 1382-1386.

- ''Osteosarcoma-Induction in the Beagle Dog with Alpha-Emitting Radionuclides; (a) Further Validation of the Linear Hypothesis of Radiation Carcinogenesis, (b) Absence of Any Suggestion of Safe Radiation Threshold for Bone Cancer Induction'', supplementary testimony to the JCAE, February 18, 1970; pages 1452-1464.
- ''The Cancer-Leukemia Risk from FRC Guideline Radiation Based upon ICRP Publications; Complete Consistency with Gofman-Tamplin Estimates'', supplementary testimony to the JCAE, February 20, 1970; pages 1465-1475.
- ''Allowable Occupational Exposures and Employee's Compensation'', supplementary testimony to the JCAE, March 30, 1970; pages 1476-1481.
- ''The History of Erroneous Handling of the Radiation Hazard Problem in Atomic Energy Development'', presented to a Congressional Seminar, April 7-8, 1970; pages 1482-1500. Non-technical.
- ''A Proposal for a Rational Future Protection Policy with Respect to Radioactivity and Other Forms of Pollution'', presented to a Congressional Seminar, April 7-8, 1970; pages 1501-1508. Non-technical.
- ''Can We Survive the Peaceful Atom?'', presented at the University of Minnesota, ''Earth Day'', April 22, 1970; pages 1509-1522. This non-technical presentation is available also in **Earth Day—The Beginning; A Guide for Survival**, compiled and edited by the National Staff of Environmental Action. A Bantam Book 553 05822 125. New York: Arno Press, Inc., May 1970.
- ''Plutonium and Public Health'', by Donald P. Geesaman, presented at the University of Colorado, April 19, 1970; pages 1523-1537.
- ''Questions for Dr. Paul Tompkins'', (head of the Federal Radiation Council), June 29, 1970; pages 1538-1559.
- ''A Proposal for a Five-Year Moratorium on Above-Ground Nuclear Power Plants'', testimony presented to the Pennsylvania State Senate, August 20, 1970; pages 1368-1382. Non-technical.
- ''A Critique of the Use of Mouse Genetic Data in Estimation of the Hazard of Radiation to Humans (Both Somatic and Genetic)'', November 12, 1970; pages 1617-1625.

□ □ □ □ □

Index by Alphabet

See also the *Index by Issues*, pages 6-14.

The Cartoonists and Artists

Book design, layout, paste-up, editing, and indexes by Egan O'Connor.

I wish to state, also, that humanity is in the debt of a New Jersey citizen who seldom receives the credit he deserves for starting the anti-nuclear movement and for tirelessly pursuing his mission virtually alone for years as, one by one, he inspired others to become concerned: Larry Bogart. *JWG.*

NUKOSAUR